PRAISE FOR *MACHINES*

"Mr. Markoff focuses on the personalities, since technology depends on the values of its creators. The human element makes the subject accessible. (His chapter on the history of AI is superb.)" —*The Economist*

"Neither alarmist nor affirmative, [*Machines of Loving Grace* contains] urgent, compelling and relevant calls to consciously embed our values in the systems we design, and to critically engage with our choices. . . . Before welcoming our robotic overlords, read [this book]." —*New Scientist*

"John Markoff of the *New York Times* highlights the compelling contrast between AI and intelligence amplification (IA). He chronicles the fascinating and often antagonistic evolution of these fields since 1956, when both terms were coined." —*Nature*

"Markoff did his homework and capably tackles interesting things." —*San Francisco Chronicle*

"Fascinating, informative, thought-provoking." —*San Jose Mercury News*

"A detailed, engrossing history of robotics. . . . This revealing look at profound technological and economic developments will unsettle anyone who has a job to lose." —*Publishers Weekly*

"Readers who like their history with a little personality will enjoy this detailed exploration of the development of computers and robotics as assistive or control technologies and the people who make it happen." —*Library Journal*

"Will robots of the future be our partners or our Frankenstein's monster? You should read this book. As Markoff explains in this engrossing narrative filled with colorful characters and head-snapping insights, the answer is up to us."

—Walter Isaacson, author of *Steve Jobs* and *The Innovators*

"How should we balance what machines can do for us, and what they can help us do ourselves? Markoff hits on one of the central questions in technology today. A fascinating read."

—Tony Fadell, CEO of Nest

"There have been books about robots ever since Karel Capek launched the trend. *Machines of Loving Grace* is the first comprehensive study to place the subject in the context of the cloud-based intelligence that throws a game-changer at the question 'But what will they do for brains?' If you want to know—or, if you don't want to know—how the servers are becoming our masters, John Markoff will help you collect your remaining thoughts."

—George Dyson, author of *Turing's Cathedral: The Origins of the Digital Universe*

"I devoured this book like an espionage thriller because the fate of humanity is on the line, and Markoff's narrative is so engaging." —Oren Etzioni, CEO of the Allen Institute for Artificial Intelligence

"John Markoff has been seeing around the corners of the technology revolution throughout his career. Now he uses his full range of vision and experience to examine whether humans can make peace with the coming wave of smart machines. His view is intelligent, illuminating, and, yes, optimistic."

—John Hollar, president and CEO of the Computer History Museum

MACHINES
OF
LOVING GRACE

ALSO BY JOHN MARKOFF

What the Dormouse Said:
How the Sixties Counterculture Shaped
the Personal Computer Industry

Takedown:
The Pursuit and Capture of Kevin Mitnick,
America's Most Wanted Computer Outlaw—
By the Man Who Did It (with Tsutomu Shimomura)

Cyberpunk:
Outlaws and Hackers on the Computer Frontier
(with Katie Hafner)

The High Cost of High Tech:
The Dark Side of the Chip (with Lenny Siegel)

MACHINES OF LOVING GRACE

THE QUEST FOR COMMON GROUND BETWEEN HUMANS AND ROBOTS

JOHN MARKOFF

An Imprint of *HarperCollins*Publishers

MACHINES OF LOVING GRACE. Copyright © 2015 by John Markoff. All rights reserved. Printed in the United States of America. No part of this book may be used or reproduced in any manner whatsoever without written permission except in the case of brief quotations embodied in critical articles and reviews. For information address HarperCollins Publishers, 195 Broadway, New York, NY 10007.

HarperCollins books may be purchased for educational, business, or sales promotional use. For information please e-mail the Special Markets Department at SPsales@harpercollins.com.

A hardcover edition of this book was published in 2015 by Ecco, an imprint of HarperCollins Publishers.

FIRST ECCO PAPERBACK EDITION PUBLISHED 2016.

Designed by Suet Yee Chong

Library of Congress Cataloging-in-Publication Data has been applied for.

ISBN 978-0-06-226669-9

16 17 18 19 20   OV/RRD   10 9 8 7 6 5 4 3 2 1

I like to think
(it has to be!)
of a cybernetic ecology
where we are free of our labors
and joined back to nature,
returned to our mammal
brothers and sisters,
and all watched over
by machines of loving grace.

—Richard Brautigan,
  "All Watched Over by Machines of Loving Grace"

CONTENTS

PREFACE

In the spring of 2014 I parked in front of the small café adjacent to the Stanford Golf Course. As I got out of my car, a woman pulled her Tesla into the next space, got out, and unloaded her golf cart. She then turned and walked toward the golf course and her golf cart followed her—on its own. I was stunned, but when I feverishly Googled "robot golf carts" I found that there was nothing new about the caddy. The Caddy-Trek robot golf cart, which retails at $1,795, is simply one of many luxury items that you might find on a Silicon Valley golf course these days.

Robots are pervading our daily lives. Cheap sensors, powerful computers, and artificial intelligence software will ensure they will, increasingly, be autonomous. They will assist us and they will replace us. They will transform health care and elder care as they have transformed warfare. Yet despite the fact that these machines have been a part of our literature and cinema for decades, we are ill-prepared for this new world now in the making.

The idea that led to this book has its roots in the years between 1999 and 2001, when I was conducting a series of interviews that would ultimately become the book *What the Dormouse Said: How the Sixties Counterculture Shaped the Personal Computer Industry*. My original research was an exercise in "anti-autobiography." I grew up in Palo Alto—a city that would become the heart of Silicon Valley—in the 1950s and the first half of the 1960s, but I moved away during a crucial decade when a set of computing and communications technologies combined to lay the foundation for personal computing and the modern Internet. I returned just in time to see the emergence of a computing era that would soon sweep through the entire world, transforming everything it touched. Years later while I was doing research for *Dormouse* I noted a striking contrast in the intent of the designers of the original interactive computer systems. At the outset of the Information Age two researchers independently set out to invent the future of computing. They established research laboratories roughly equidistant from the Stanford University campus. In 1964 John McCarthy, a mathematician and computer scientist who had coined the term "artificial intelligence," began designing a set of technologies that were intended to simulate human capabilities, a project he believed could be completed in just a decade. At the same time, on the other side of campus, Douglas Engelbart, who was a dreamer intent on using his expertise to improve the world, believed that computers should be used to "augment" or extend human capabilities, rather than to mimic or replace them. He set out to create a system to permit small groups of knowledge workers to quickly amplify their intellectual powers and work collaboratively. One researcher attempted to replace human beings with intelligent machines, while the other aimed to extend human capabilities. Of course, together, their work defined both a dichotomy and a paradox. The paradox is that the same technologies that extend the intellectual power of humans can displace them as well.

In this book, I have attempted to capture the ways in which scientists, engineers, and hackers have grappled with questions about the deepening relationship between human and machine. In some cases I discovered that the designers resist thinking deeply about the paradoxical relationship between artificial intelligence and intelligence augmentation. Often, it comes down to a simple matter of economics. There is now a burgeoning demand for robots with abilities that far exceed those of the early industrial robots of the last half century. Even in already highly automated industries like agriculture, a new wave of "ag robots" are now driving tractors and harvesters, irrigating and weeding fields, conducting surveillance from the air, and generally increasing farm productivity.

There are also many instances where the researchers think deeply about the paradox, and many of those researchers place themselves squarely in Engelbart's camp. Eric Horvitz, for example, is a Microsoft artificial intelligence researcher, medical doctor, and past president of the Association for the Advancement of Artificial Intelligence, who has for decades worked on systems to extend human capabilities in the office. He has designed elaborate robots that serve as office secretaries, performing tasks like tracking calendars, greeting visitors, and managing interruptions and distractions. He is building machines that will simultaneously augment and displace humans.

Others, like German-born Sebastian Thrun, an artificial intelligence researcher and a roboticist (also a cofounder at Udacity, an online education company) are building a world that will be full of autonomous machines. As founder of the Google car project, Thrun led the design of autonomous vehicle technology that one day may displace millions of human drivers, something he justifies by citing the lives it will save and the injuries it will avoid.

The central topic of this book is the dichotomy and the paradox inherent in the work of the designers who alternatively augment and replace humans in the systems they build. This distinction is clearest in the contrasting philosophies of Andy Rubin and Tom Gruber. Rubin was the original architect of Google's robot empire and Gruber is a key designer of Apple's Siri intelligent assistant. They are both among Silicon Valley's best and brightest, and their work builds on that of their predecessors—Rubin mirrors John McCarthy and Gruber follows Doug Engelbart—to alternatively replace or augment humans.

Today, both robotics and artificial intelligence software increasingly evoke memories of the early days of personal computing. Like the hobbyists who created the personal computer industry, AI designers and roboticists are hugely enthusiastic about the technological advances, new products, and companies clearly ahead of them. At the same time, many of the software designers and robot engineers grow uncomfortable when asked about the potential consequences of their inventions and frequently deflect questions with gallows humor. Yet questions are essential. There's no blind watchmaker for the evolution of machines. Whether we augment or automate is a design decision that will be made by individual human designers.

It would be easy to cast one group as heroes and the other as villains, yet the consequences are too nuanced to be easily sorted into black and white categories. Between their twin visions of artificial intelligence and robotics lies a future that might move toward a utopia, a dystopia, or somewhere in between. Is an improved standard of living and relief from drudgery worthwhile if it also means giving up freedom and privacy? Is there a right or a wrong way to design these systems? The answer, I believe, lies with the designers themselves. One group designs powerful machines that allow humans to perform previously unthinkable tasks, like programming robots for space exploration, while the other works to replace humans with machines, like the developers of artificial intelligence software that enables robots to perform the work of doctors and lawyers. It is essential that these two camps find a way to communicate with each other. How we design and interact with our increasingly autonomous machines will determine the nature of our society and our economy. It will increasingly determine every aspect of our modern world, from whether we live in a more or less stratified society to what it will mean to be human.

The United States is currently in the midst of a renewed debate about the consequences of artificial intelligence and robotics and their impact on both employment and the quality of life. It is a strange time—workplace automation has started to strike the white-collar workforce with the same ferocity that it transformed the factory floor beginning in the 1950s. Yet the return of the "great automation debate" a half century after the initial one feels sometimes like scenes from *Rashomon:* everyone sees the same story but interprets it in a different, self-serving way. Despite ever-louder warnings about the dire consequences of computerization, the number of Americans in the workforce has continued to grow. Analysts look at the same Bureau of Labor Statistics data and simultaneously predict both the end of work and a new labor renaissance. Whether labor is vanishing or being transformed, it's clear that this new automation age is having a profound impact on society. Less clear, despite vast amounts both said and written, is whether anyone truly grasps where technological society is headed.

Although few people encountered the hulking mainframe computers of the 1950s and 1960s, there was a prevailing sense that these machines exerted some sinister measure of control over their lives. Then in the 1970s personal computing arrived and the computer became something much friendlier—because people could touch these computers, they began to feel that they were now in control. Today, an "Internet of Things" is emerging and computers have once again started to "disappear," this time blending into everyday objects that have as a result acquired seemingly magical powers—our smoke detectors speak and listen to us. Our phones, music players, and tablets have more computing power than the supercomputers of just a few decades ago.

With the arrival of "ubiquitous computing," we have entered a new age of smart machines. In the coming years, artificial intelligence and robotics will have an impact on the world more dramatic than the changes personal computing and the Internet have brought in the past three decades. Cars will drive themselves and robots will do the work of FedEx employees and, inevitably, doctors and lawyers. The new era offers the promise of great physical and computing power, but it also reframes the question first raised more than fifty years ago: Will we control these systems or will they control us?

George Orwell posed the question eloquently. *1984* is remembered for its description of the Surveillance State, but Orwell also wrote about the idea that state control would be exercised by shrinking human spoken and written language to make it more difficult to express, and thus conceive of, alternative ideas. He posited a fictional language, "Newspeak," that effectively limited freedom of thought and self-expression.

With the Internet offering millions of channels, at first glance we couldn't be farther today from Orwell's nightmare, but in a growing number of cases, smart machines are making decisions for us. If these systems merely offered advice, we could hardly call these interactions "controlling" in an Orwellian sense. However, the much-celebrated world of "big data" has resulted in a vastly different Internet from the one that existed just a decade ago. The Internet has extended the reach of computing and is transforming our culture. This neo-Orwellian society presents a softer form of control. The Internet offers unparalleled new freedoms while paradoxically extending control and surveillance far beyond what Orwell originally conceived. Every footstep and every utterance is now tracked and collected, if not by Big Brother then by a growing array of commercial "Little Brothers." The Internet has become an intimate technology that touches every facet of our culture. Today our smartphones, laptops, and desktop computers listen to us, supposedly at our command, and cameras gaze from their screens as well, perhaps benignly. The impending Internet of Things is now introducing unobtrusive, always-on, and supposedly helpful countertop robots, like the Amazon Echo and Cynthia Breazeal's Jibo, to homes across the country.

Will a world that is watched over by what sixties poet Richard Brautigan described as "machines of loving grace" be a free world? Free, that is, in the sense of "freedom of speech," rather than "free beer."[1] The best way to answer questions about control in a world full of smart machines is by understanding the values of those who are actually building these systems.

In Silicon Valley it is popular for optimistic technologists to believe that the twin forces of innovation and Moore's law—the doubling of computing power at two-year intervals—are sufficient to account for technical progress. Little thought is given as to why one technology wins out over others, or why a particular technology arises when it does. This view is anathema to what social scientists call the "social construction of technology"—the understanding that we shape our tools rather than being shaped by them.

We have centuries of experience with machines such as the backhoe and steam shovel, both of which replace physical labor. Smart machines that displace white-collar workers and intellectual labor, however, are a new phenomenon. More than merely replacing humans, information technology is democratizing certain experiences. It is not just that using a personal computer has made it possible to dispense with a secretary. The Internet and the Web have vastly reduced the costs of journalism, for example, not just upending the newspaper industry but also fundamentally transforming the process of collecting and reporting the news. Similarly technologies like pitch correction have made it possible for anyone to sing on key without training while a variety of computerized music systems allow anyone to become a composer and a musician. In the future, how these systems are designed will foretell either a great renaissance or possibly something darker—a world in which human skills are passed on wholesale to machines. McCarthy's and Engelbart's work defined a new era in which digital computers would transform economies and societies as profoundly as did the industrial revolution.

Recent experiments that guaranteed a "basic income" in the poorest part of the world may also offer a profound insight into the future of work in the face of encroaching, brilliant machines. The results of these experiments were striking because they ran counter to the popular idea that economic security undercuts the will to work. An experiment in an impoverished village in India in 2013 guaranteeing basic needs had just the opposite effect. The poor did not rest easy on their government subsidies; instead, they became more responsible and productive. It is quite likely that we will soon have the opportunity to conduct a parallel experiment in the First World. The idea of a basic income is already on the political agenda in Europe. Raised by the Nixon administration in the form of a negative income tax in 1969, the idea is currently not politically acceptable in the United States. However, that will change quickly if technological unemployment becomes widespread.

What will happen if our labor is no longer needed? If jobs for warehouse workers, garbage collectors, doctors, lawyers, and journalists are displaced by technology? It is of course impossible to know this future, but I suspect society will find that humans are hardwired to work or find an equivalent way to produce something of value in the future. A new economy will create jobs that we are unable to conceive of today. Science-fiction writers, of course, have already covered this ground well. Read John Barnes's *Mother of Storms* or Charlie Stross's *Accelerando* for a compelling window into what a future economy might look like. The simple answer is that human creativity is limitless, and if our basic needs are looked after by robots and AIs, we will find ways to entertain, educate, and care for one another in new ways. The answers may be murky but the questions are increasingly sharp. Will these intelligent machines that interact with and care for us be our allies or will they enslave us?

In the pages that follow I portray a diverse set of computer scientists, hackers, roboticists, and neuroscientists. They share a growing sense that we are approaching an inflection point where humans will live in a world of machines that mimic, and even surpass, some human capabilities. They offer a rainbow of sensibilities about our place in this new world.

During the first half of this century, society will be tasked with making hard decisions about the smart machines that have the potential to be our servants, partners, or masters. At the very dawn of the computer era in the middle of the last century, Norbert Wiener issued a warning about the potential of automation: "We can be humble and live a good life with the aid of the machines," he wrote, "or we can be arrogant and die."

It is still a fair warning.

John Markoff
San Francisco, California
January 2015

1 | BETWEEN HUMAN AND MACHINE

Bill Duvall was already a computer hacker when he dropped out of college. Not long afterward he found himself face-to-face with Shakey, a six-foot-tall wheeled robot. Shakey would have its moment in the sun in 1970 when *Life* magazine dubbed it the first "electronic person." As a robot, Shakey fell more into the R2-D2 category of mobile robots than the more humanoid C-3PO of Star Wars lore. It was basically a stack of electronic gear equipped with sensors and motorized wheels, first tethered, then later wirelessly connected to a nearby mainframe computer.

Shakey wasn't the world's first mobile robot, but it was the first one that was designed to be truly autonomous. An early experiment in artificial intelligence (AI), Shakey was intended to reason about the world around it, plan its own actions, and perform tasks. It could find and push objects and move around in a planned way in its highly structured world. Moreover, as a harbinger of things to come, it was a prototype for much more ambitious machines that were intended to live, in military parlance, in "a hostile environment."

Although the project has now largely been forgotten, the Shakey designers pioneered computing technologies today used by more than one billion people. The mapping software in everything from cars to smartphones is based on techniques that were first developed by the Shakey team. Their A* algorithm is the best-known way to find the shortest path between two locations. Toward the end of the project, speech control was added as a research task, and today Apple's Siri speech service is a distant descendant of the machine that began life as a stack of rolling actuators and sensors.

Duvall had grown up on the Peninsula south of San Francisco, the son of a physicist who was involved in classified research at Stanford Research Institute, the military-oriented think tank where Shakey resided. At UC Berkeley he took all the computer programming courses the university offered in the mid-1960s. After two years he dropped out to join the think tank where his father worked, just miles from the Stanford campus, entering a cloistered priesthood where the mainframe computer was the equivalent of a primitive god.

For the young computer hacker, Stanford Research Institute, soon after renamed SRI International, was an entry point into a world that allowed skilled programmers to create elegant and elaborate software machines. During the 1950s SRI pioneered the first check-processing computers. Duvall arrived to work on an SRI contract to automate an English bank's operations, but the bank had been merged into a larger bank, and the project was put on an indefinite hold. He used the time for his first European vacation and then headed back to Menlo Park to renew his romance with computing, joining the team of artificial intelligence researchers building Shakey.

Like many hackers, Duvall was something of a loner. In high school, a decade before the movie *Breaking Away*, he joined a local cycling club and rode his bike in the hills behind

Stanford. In the 1970s the movie would transform the American perception of bike racing, but in the 1960s cycling was still a bohemian sport, attracting a ragtag assortment of individualists, loners, and outsiders. That image fit Duvall's worldview well. Before high school he attended the Peninsula School, an alternative elementary and middle school that adhered to the philosophy that children should learn by doing and at their own pace. One of his teachers had been Ira Sandperl, a Gandhi scholar who was a permanent fixture behind the cash register at Kepler's, a bookstore near the Stanford Campus. Sandperl had also been Joan Baez's mentor and had imbued Duvall with an independent take on knowledge, learning, and the world.

Duvall was one of the first generation of computer hackers, a small subculture that had originally emerged at MIT, where computing was an end in itself and where the knowledge and code needed to animate the machines were both freely shared. The culture had quickly spread to the West Coast, where it had taken root at computing design centers like Stanford and the University of California at Berkeley.

It was an era in which computers were impossibly rare—a few giant machines were hidden away in banks, universities, and government-funded research centers. At SRI, Duvall had unfettered access to a room-sized machine first acquired for an elite military-funded project and then used to run the software controlling Shakey. At both SRI and at the nearby Stanford Artificial Intelligence Laboratory (SAIL), tucked away in the hills behind Stanford University, there was a tightly knit group of researchers who already believed in the possibility of building a machine that mimicked human capabilities. To this group, Shakey was a striking portent of the future, and they believed that the scientific breakthrough to enable machines to act like humans would come in just a few short years.

Indeed, during the mid-sixties there was virtually boundless optimism among the small community of artificial intelligence researchers on both coasts. In 1966, when SRI and SAIL were beginning to build robots and AI programs in California, another artificial intelligence pioneer, Marvin Minsky, assigned an undergraduate to work on the problem of computer vision on the other side of the country, at MIT. He envisioned it as a summer project. The reality was disappointing. Although AI might be destined to transform the world, Duvall, who worked on several SRI projects before transferring to the Shakey project to work in the trenches as a young programmer, immediately saw that the robot was barely taking baby steps.

Shakey lived in a large open room with linoleum floors and a couple of racks of electronics. Boxlike objects were scattered around for the robot to "play" with. The mainframe computer providing the intelligence was nearby. Shakey's sensors would capture the world around it and then "think"—standing motionless for minutes on end—before resuming its journey, even in its closed and controlled world. It was like watching grass grow. Moreover, it frequently broke down or would drain its batteries after just minutes of operation.

For a few months Duvall made the most of his situation. He could see that the project was light-years away from the stated goal of an automated military sentry or reconnaissance agent. He tried to amuse himself by programming the rangefinder, a clunky device based on a rotating mirror. Unfortunately it was prone to mechanical failure, making software development a highly unsatisfying exercise in error prediction and recovery. One of the managers told him that the project was in need of a "probabilistic decision tree" to refine the robot's vision system. So rather than working on that special-purpose mechanism, he spent his time writing a programming tool that could generate such trees programmatically. Shakey's vision system worked better than the rangefinder. Even with the simplest machine vision processing, it could identify both edges and basic shapes, essential primitives to understand and travel in its surroundings.

Duvall's manager believed in structuring his team so that

"science" would only be done by "scientists." Programmers were low-status grunt workers who implemented the design ideas of their superiors. While some of the leaders of the group appeared to have a high-level vision to pursue, the project was organized in a military fashion, making work uninspiring for a low-level programmer like Duvall, stuck writing device drivers and other software interfaces. That didn't sit well with the young computer hacker.

Robots seemed like a cool idea to him, but before Star Wars there weren't a lot of inspiring models. There was Robby the Robot from *Forbidden Planet* in the 1950s, but it was hard to find inspiration in a broader vision. Shakey simply didn't work very well. Fortunately Stanford Research Institute was a big place and Duvall was soon attracted by a more intriguing project.

Just down the hall from the Shakey laboratory he would frequently encounter another research group that was building a computer to run a program called NLS, the oN-Line System. While Shakey was managed hierarchically, the group run by computer scientist Doug Engelbart was anything but. Engelbart's researchers, an eclectic collection of buttoned-down white-shirted engineers and long-haired computer hackers, were taking computing in a direction so different it was not even in the same coordinate system. The Shakey project was struggling to mimic

Artificial intelligence pioneer Charles Rosen with Shakey, the first autonomous robot. The Pentagon funded the project to research the idea of a future robotic sentry. (*Image courtesy of SRI International*)

the human mind and body. Engelbart had a very different goal. During World War II he had stumbled across an article by Vannevar Bush, who had proposed a microfiche-based information retrieval system called Memex to manage all of the world's knowledge. Engelbart later decided that such a system could be assembled based on the then newly available computers. He thought the time was right to build an interactive system to capture knowledge and organize information in such a way that it would now be possible for a small group of people—scientists, engineers, educators—to create and collaborate more effectively. By this time Engelbart had already invented the computer mouse as a control device and had also conceived of the idea of hypertext links that would decades later become the foundation for the modern World Wide Web. Moreover, like Duvall, he was an outsider within the insular computer science world that worshipped theory and abstraction as fundamental to science.

The cultural gulf between the worlds defined by artificial intelligence and Engelbart's contrarian idea, deemed "intelligence augmentation"—he referred to it as "IA"—was already palpable. Indeed, when Engelbart paid a visit to MIT during the 1960s to demonstrate his project, Marvin Minsky complained that it was a waste of research dollars on something that would create nothing more than a glorified word processor.

Despite earning no respect from establishment computer scientists, Engelbart was comfortable with being viewed as outside the mainstream academic world. When attending the Pentagon DARPA review meetings that were held regularly to bring funded researchers together to share their work, he would always begin his presentations by saying, "This is not computer science." And then he would go on to sketch a vision of using computers to permit people to "bootstrap" their projects by making learning and innovation more powerful.

Even if it wasn't in the mainstream of computer science, the ideas captivated Bill Duvall. Before long he switched his allegiance and moved down the hall to work in Engelbart's lab. In the space of less than a year he went from struggling to program the first useful robot to writing the software code for the two computers that first connected over a network to demonstrate what would evolve to become the Internet. Late in the evening on October 29, 1969, Duvall connected Engelbart's NLS software in Menlo Park to a computer in Los Angeles controlled by another young hacker via a data line leased from the phone company. Bill Duvall would become the first to make the leap from research to replace humans with computers to using computing to augment the human intellect, and one of the first to stand on both sides of an invisible line that even today divides two rival, insular engineering communities.

Significantly, what started in the 1960s was then accelerated in the 1970s at a third laboratory also located near Stanford. Xerox's Palo Alto Research Center extended ideas originally incubated at McCarthy's and Engelbart's labs, in the form of the personal computer and computer networking, which were in turn successfully commercialized by Apple and Microsoft. Among other things, the personal computing industry touched off what venture capitalist John Doerr identified during the 1990s as the "largest legal accumulation of wealth in history."[1]

Most people know Doug Engelbart as the inventor of the mouse, but his more encompassing idea was to use a set of computer technologies to make it possible for small groups to "bootstrap" their projects by employing an array of ever more powerful software tools to organize their activities, creating what he described as the "collective IQ" that outstripped the capabilities of any single individual. The mouse was simply a gadget to improve our ability to interact with computers.

In creating SAIL, McCarthy's impact upon the world was equal to Engelbart's in many ways. People like Alan Kay and

Larry Tesler, who were both instrumental in the design of the modern personal computer, passed through his lab on their way to Xerox and subsequently to Apple Computer. Whitfield Diffie took away ideas that would lead to the cryptographic technology that secures modern electronic commerce.

There were, however, two other technologies being developed simultaneously at SRI and SAIL that are only now beginning to have a substantial impact: robotics and artificial intelligence software. Both of these are not only in the process of transforming economies; they are fostering a new era of intelligent machines that is fundamentally changing the way we live.

The impact of both computing and robotics had been forecast before these laboratories were established. Norbert Wiener invented the concept of cybernetics at the very dawn of the computing era in 1948. In his book *Cybernetics,* he outlined a new engineering science of control and communication that foreshadowed both technologies. He also foresaw the implications of these new engineering disciplines, and two years after he wrote *Cybernetics,* his companion book, *The Human Use of Human Beings,* explored both the value and the danger of automation.

He was one of the first to foresee the twin possibilities that information technology might both escape human control and come to control human beings. More significantly he posed an early critique of the arrival of machine intelligence: the danger of passing decisions on to systems that, incapable of thinking abstractly, would make decisions in purely utilitarian terms rather than in consideration of richer human values.

Engelbart worked as an electronics technician at NASA's Ames Research Center during the 1950s, and he had watched as aeronautical engineers first built small models to test in a wind tunnel and then scaled them up into full-sized airplanes. He quickly realized that the new silicon computer circuits could be scaled in the opposite direction—down into what would become known as the "microcosm." By shrinking the circuitry it would be possible to place more circuits in the same space for the same cost. And dramatically, each time the circuit density increased, performance improvement would not be additive, but rather multiplicative. For Engelbart, this was a crucial insight. Within a year after the invention of the modern computer chip in the late 1950s he understood that there would ultimately be enough cheap and plentiful computing power to change the face of humanity.

This notion of exponential change—Moore's law, for example—is one of the fundamental contributions of Silicon Valley. Computers, Engelbart and Moore saw, would become more powerful ever more quickly. Equally dramatically, their cost would continue falling, not incrementally, but also at an accelerating rate, to the point where soon remarkably powerful computers would be affordable by even the world's poorest people. During the past half decade that acceleration has led to rapid improvement in technologies that are necessary components for artificial intelligence: computer vision, speech recognition, and robotic touch and manipulation. Machines now also taste and smell, but recently more significant innovations have come from modeling human neurons in electronic circuits, which has begun to yield advances in pattern recognition—mimicking human cognition.

The quickening pace of AI innovation has led some, such as Rice University computer scientist Moshe Vardi, to proclaim the imminent end of a very significant fraction of all tasks performed by humans, perhaps as soon as 2045.[2] Even more radical voices argue that computers are evolving at such a rapid pace that they will outstrip the intellectual capabilities of humans in one, or at the most two more generations. The science-fiction author and computer scientist Vernor Vinge posed the notion of a computing "singularity" in which machine intelligence will make such rapid progress that it will cross a threshold and then in some as yet unspecified leap, become superhuman.

It is a provocative claim, but far too early to answer definitively. Indeed, it is worthwhile recalling the point made by long-time Silicon Valley observer Paul Saffo when thinking about the compounding impact of computing. "Never mistake a clear view for a short distance," he has frequently reminded the Valley's digerati. For those who believe that human labor will be obsolete in the space of a few decades, it's worth remembering that even against the background of globalization and automation, between 1980 and 2010, the U.S. labor force actually continued to expand. Economists Frank Levy and Richard J. Murnane recently pointed out that since 1964 the economy has actually added seventy-four million jobs.[3]

MIT economist David Autor has offered a detailed explanation of the consequences of the current wave of automation. Job destruction is not across the board, he argues, but instead has focused on the routinized tasks performed by those in the middle of the job structure—the post–World War II white-collar expansion. The economy has continued to expand at both the bottom and the top of the pyramid, leaving the middle class vulnerable while expanding markets for both menial and expert jobs.

Rather than extending that debate here, however, I am interested in exploring a different question first posed by Norbert Wiener in his early alarms about the introduction of automation. What will the outcome of McCarthy's and Engelbart's differing approaches be? What are the consequences of the design decisions made by today's artificial intelligence researchers and roboticists, who, with ever greater ease, can choose between extending and replacing the "human in the loop" in the systems and products they create? By the same token, what are the social consequences of building intelligent systems that substitute for or interact with humans in business, entertainment, and day-to-day activities?

Two distinct technical communities with separate traditions, values, and priorities have emerged in the computing world. One, artificial intelligence, has relentlessly pressed ahead toward the goal of automating the human experience. The other, the field of human-computer interaction, or HCI, has been more concerned with the evolution of the idea of "man-machine symbiosis" that was foreseen by pioneering psychologist J. C. R. Licklider at the dawn of the modern computing era as an interim step on the way to brilliant machines. Significantly, Licklider, as director of DARPA's Information Processing Techniques Office in the mid-1960s, would be an early funder of both McCarthy and Engelbart. It was the Licklider era that would come to define the period when the Pentagon agency operated as a truly "blue-sky" funding organization, a period when, many argue, the agency had its most dramatic impact.

Wiener had raised an early alert about the relationship between man and computing machines. A decade later Licklider pointed to the significance of the impending widespread use of computing and how the arrival of computing machines was different from the previous era of industrialization. In a darker sense Licklider also forecast the arrival of the Borg of *Star Trek* notoriety. The Borg, which entered popular culture in 1988, was a proposed cybernetic alien species that assembles into a "hive mind" in which the collective subsumes the individual, intoning the phrase, "You will be assimilated."

Licklider wrote in 1960 about the distance between "mechanically extended man" and "artificial intelligence," and warned about the early direction of automation technology: "If we focus upon the human operator within the system, however, we see that, in some areas of technology, a fantastic change has taken place during the last few years. 'Mechanical extension' has given way to replacement of men, to automation, and the men who remain are there more to help than to be helped. In some instances, particularly in large computer-centered information and control systems, the human operators are responsible mainly for functions that it proved infeasible to automate."[4] That observation seems fatalistic in accepting the shift toward automation rather than augmentation.

Licklider, like McCarthy a half decade later, was confident that the advent of "Strong" artificial intelligence—a machine capable of at least matching wits and self-awareness with a human—was likely to arrive relatively soon. The period of man-machine "symbiosis" might only last for less than two decades, he wrote, although he allowed that the arrival of truly smart machines that were capable of rivaling thinking humans might not happen for a decade, or perhaps fifty years.

Ultimately, although he posed the question of whether humans will be freed or enslaved by the Information Age, he chose not to directly address it. Instead he drew a picture of what has become known as a "cyborg"—part human, part machine. In Licklider's view human operators and computing equipment would blend together seamlessly to become a single entity. That vision has since been both celebrated and reviled. But it still begs the unanswered question—will we be masters, slaves, or partners of the intelligent machines that are appearing today?

Consider the complete spectrum of human-machine interactions from simple "FAQbots" to Google Now and Apple's Siri. Moving into the unspecified future in the movie *Her,* we see an artificial intelligence, voiced by Scarlett Johansson, capable of carrying on hundreds of simultaneous, intimate, human-level conversations. Google Now and Siri currently represent two dramatically different computer-human interaction styles. While Siri intentionally and successfully mimics a human, complete with a wry sense of humor, Google Now opts instead to function as a pure information oracle, devoid of personality or humanity.

It is tempting to see the personalities of the two competing corporate chieftains in these contrasting approaches. At

Apple, Steve Jobs saw the potential in Siri before it was even capable of recognizing human speech and focused his designers on natural language as a better way to control a computer. At Google, Larry Page, by way of contrast, has resisted portraying a computer in human form.

How far will this trend go? Today it is anything but certain. Although we are already able to chatter with our cars and other appliances using limited vocabularies, computer speech and voice understanding is still a niche in the world of "interfaces" that control the computers that surround us. Speech recognition clearly offers a dramatic improvement in busy-hand, busy-eye scenarios for interacting with the multiplicity of Web services and smartphone applications that have emerged. Perhaps advances in brain-computer interfaces will prove to be useful for those unable to speak or when silence or stealth is needed, such as card counting in blackjack. The murkier question is whether these cybernetic assistants will eventually pass the Turing test, the metric first proposed by mathematician and computer scientist Alan Turing to determine if a computer is "intelligent." Turing's original 1951 paper has spawned a long-running philosophical discussion and even an annual contest, but today what is more interesting than the question of machine intelligence is what the test implies about the relationship between humans and machines.

Turing's test consisted of placing a human before a computer terminal to interact with an unknown entity through typewritten questions and answers. If, after a reasonable period, the questioner was unable to determine whether he or she was communicating with a human or a machine, then the machine could be said to be "intelligent." Although it has several variants and has been widely criticized, from a sociological point of view the test poses the right question. In other words, it is relevant with respect to the human, not the machine.

In the fall of 1991 I covered the first of a series of Turing test contests sponsored by a New York City philanthropist,

Hugh Loebner. The event was first held at the Boston Computer Museum and attracted a crowd of computer scientists and a smattering of philosophers. At that point the "bots," software robots designed to participate in the contest, weren't very far advanced beyond the legendary Eliza program written by computer scientist Joseph Weizenbaum during the 1960s. Weizenbaum's program mimicked a Rogerian psychologist (a human-centered form of psychiatry focused on persuading a patient to talk his or her way toward understanding his or her actual feelings) and he was horrified to discover that his students had become deeply immersed in intimate conversations with his first, simple bot.

But the judges for the original Loebner contest in 1991 fell into two broad categories: computer literate and computer illiterate. For human judges without computer expertise, it turned out that for all practical purposes the Turing test was conquered in that first year. In reporting on the contest I quoted one of the nontechnical judges, a part-time auto mechanic, saying why she was fooled: "It typed something that I thought was trite, and when I responded it interacted with me in a very convincing fashion,"[5] she said. It was a harbinger of things to come. We now routinely interact with machines simulating humans and they will continue to improve in convincing us of their faux humanity.

Today, programs like Siri not only seem almost human; they are beginning to make human-machine interactions in natural language seem routine. The evolution of these software robots is aided by the fact that humans appear to want to believe they are interacting with humans even when they are conversing with machines. We are hardwired for social interaction. Whether or not robots move around to assist us in the physical world, they are already moving among us in cyberspace. It's now inevitable that these software bots—AIs, if only of limited capability—will increasingly become a routine part of daily life.

Intelligent software agents such as Apple's Siri, Microsoft's Cortana, and Google Now are interacting with hundreds of millions of people, by default defining this robot/human relationship. Even at this relatively early stage Siri has a distinctly human style, a first step toward the creation of a generation of likable and trusted advisors. Will it matter whether we interact with these systems as partners or keep them as slaves? While there is an increasingly lively discussion about whether intelligent agents and robots will be autonomous—and if they are autonomous, whether they will be self-aware enough that we need to consider questions of "robot rights"—in the short term the more significant question is how we treat these systems and what the design of those interactions says about what it means to be human. To the extent that we treat these systems as partners it will humanize us. Yet the question of what the relationship between humans and machines will be has largely been ignored by much of the modern computing world.

Jonathan Grudin, a computer scientist at Microsoft Research, has noted that the separate disciplines of artificial intelligence and human-computer interaction rarely speak to one another.[6] He points to John McCarthy's early explanation of the direction of artificial intelligence research: "[The goal] was to get away from studying human behavior and consider the computer as a tool for solving certain classes of problems. Thus AI was created as a branch of computer science and not as a branch of psychology."[7] McCarthy's pragmatic approach can certainly be justified by the success the field has had in the past half decade. Artificial intelligence researchers like to point out that aircraft can fly just fine without resorting to flapping their wings—an argument that asserts that to duplicate human cognition or behavior, it is not necessary to comprehend it. However, the chasm between AI and IA has only deepened as AI systems have become increasingly facile at human tasks, whether it is seeing, speaking, moving boxes, or playing chess, *Jeopardy!*, or Atari video games.

Terry Winograd was one of the first to see the two extremes

clearly and to consider the consequences. His career traces an arc from artificial intelligence to intelligence augmentation. As a graduate student at MIT in the 1960s, he focused on understanding human language in order to build a software equivalent to Shakey—a software robot capable of interacting with humans in conversation. Then, during the 1980s, in part because of his changing views on the limits of artificial intelligence, he left the field—a shift in perspective moving from AI to IA. Winograd walked away from AI in part because of a series of challenging conversations with a group of philosophers at the University of California. A member of a small group of AI researchers, he engaged in a series of weekly seminars with Berkeley philosophers Hubert Dreyfus and John Searle. The philosophers convinced him that there were real limits to the capabilities of intelligent machines. Winograd's conversion coincided with the collapse of a nascent artificial intelligence industry known as the "AI Winter." Several decades later, Winograd, who was faculty advisor for Google cofounder Larry Page at Stanford, famously counseled the young graduate student to focus on the problem of Web search rather than self-driving cars.

In the intervening decades Winograd had become acutely aware of the importance of the designer's point of view. The separation of the fields of AI and human-computer interaction, or HCI, is partly a question of approach, but it's also an ethical stance about designing humans either into or out of the systems we create. More recently at Stanford Winograd helped create an academic program focusing on "Liberation Technologies," which studies the construction of computerized systems based on human-centered values.

Throughout human history, technology has displaced human labor. Locomotives and tractors, however, didn't make human-level decisions. Increasingly, "thinking machines" will. It is also clear that technology and humanity coevolve, which again will pose the question of who will be in control. In Silicon Valley it has become fashionable to celebrate the rise of the machines, most clearly in the emergence of organizations like the Singularity Institute and in books like Kevin Kelly's 2010 *What Technology Wants*. In an earlier book in 1994, *Out of Control*, Kelly came down firmly on the side of the machines. He described a meeting between AI pioneer Marvin Minsky and Doug Engelbart:

> When the two gurus met at MIT in the 1950s, they are reputed to have had the following conversation:
>
> Minsky: We're going to make machines intelligent. We are going to make them conscious!
>
> Engelbart: You're going to do all that for the machines? What are you going to do for the people?
>
> This story is usually told by engineers working to make computers more friendly, more humane, more people centered. But I'm squarely on Minsky's side—on the side of the made. People will survive. We'll train our machines to serve us. But what are we going to do for the machines?[8]

Kelly is correct to point out that there are Minsky and Engelbart "sides." But to say that people will "survive" belittles the consequences. He is basically echoing Minsky, who is famously said to have responded to a question about the significance of the arrival of artificial intelligence by saying, "If we're lucky, maybe they'll keep us as pets."

Minsky's position is symptomatic of the chasm between the AI and IA camps. The artificial intelligence community has until now largely chosen to ignore the consequences of the systems it considers merely powerful tools, dispensing with discussions of morality. As one of the engineers who is building next-generation robots told me when I asked about the impact of automation on people: "You can't think about that; you just have to decide that you are going to do the best you can to improve the world for humanity as a whole."

During the past half century, McCarthy's and Engelbart's

philosophies have remained separate and their central conflict stands unresolved. One approach supplants humans with an increasingly powerful blend of computer hardware and software. The other extends our reach intellectually, economically, and socially using the same ingredients. While the chasm between these approaches has been little remarked, the explosion of this new wave of technology, which now influences every aspect of modern life, will encapsulate the repercussions of this divide.

Will machines supplant human workers or augment them? On one level, they will do both. But once again, that is the wrong question to ask, and it provides only a partial answer. Both software and hardware robots are flexible enough that they can ultimately become whatever we program them to be. In our current economy, how robots—both machines and intelligent systems—are designed and how they are used is overwhelmingly defined by cost and benefit, and costs are falling at an increasingly rapid rate. In our society, economics dictate that if a task can be done more cheaply by machine—software or hardware—in most cases it will be. It's just a matter of when.

The decision to come down on either side of the debates is doubly difficult because there are no obvious right answers. Although driverless cars will displace millions of jobs, they will also save many lives. Today, decisions about implementing technologies are made largely on the basis of profitability and efficiency, but there is an obvious need for a new moral calculus. The devil, however, is in more than the details. As with nuclear weapons and nuclear power, artificial intelligence, genetic engineering, and robotics will have society-wide consequences, both intended and unintended, in the next decade.

2 | A CRASH IN THE DESERT

O n a desert road near Florence, Arizona, one morning in the fall of 2005, a Volkswagen Touareg was kicking up a dust cloud, bouncing along at a steady twenty to twenty-five miles per hour, carrying four passengers. To the casual observer there was nothing unusual about the way the approaching vehicle was being driven. The road was particularly rough, undulating up and down through a landscape dotted with cactus and scrubby desert vegetation. The car bounced and wove, and all four occupants were wearing distinctive crash helmets. The Touareg was plastered with decals like a contestant in the Baja 1000 off-road race. It was also festooned with five curious sensors perched at the front of the roof, each with an unobstructed view of the road. Other sensors, including several radars, also sprouted from the roof. A video camera peered out through the windshield. A tall whip antenna set at the back of the vehicle, in combination with the sensors, conspired to give a postapocalyptic vibe reminiscent of a Mad Max movie.

The five sensors on the roof were actually mechanical contraptions, each rapidly sweeping an infrared laser beam back and forth over the road ahead. The beams, invisible to the eye, constantly reflected off the gravel road and the desert surrounding the vehicle. Bouncing back to sensors, the lasers provided a constantly changing portrait of the surrounding landscape accurate to the centimeter. Even small rocks on the road hundreds of feet ahead could not escape the unblinking gaze of the sensors, known as lidar.

The Touareg was even more peculiar inside. The driver, Sebastian Thrun, a roboticist and artificial intelligence researcher, wasn't driving. Instead he was gesturing with his hands as he chatted with the other passengers. His eyes rarely watched the road. Most striking of all: his hands never touched the steering wheel, which twitched back and forth as if controlled by some unseen ghost.

Sitting behind Thrun was another computer researcher, Mike Montemerlo, who wasn't driving either. His eyes were buried in the screen of a laptop computer that was displaying the data from the lasers, radars, and cameras in a God's-eye view of the world around the car in which potential obstacles appeared as a partial rainbow of blips on a radar screen. It revealed an ever-changing cloud of colored dots that in aggregate represented the road unfolding ahead in the desert.

The car, named Stanley, was being piloted by an ensemble of software programs running on five computers installed in the trunk. Thrun was a pioneer of an advanced version of a robotic navigation technique known as SLAM, which stands for simultaneous localization and mapping. It had become a standard tool for robots to find their way through previously unexplored terrain. The wheel continued to twitch back and forth as the car rolled along the rutted road lined with cactus and frequent outcroppings of boulders. Immediately to Thrun's right, between the front seats, was a large red E-Stop button to override the car's autopilot in an emergency. After a half-dozen miles, the robotic meanderings of the Touareg felt anticlimactic. Stanley wasn't driving down the freeway, so as the desert scenery slid by, it seemed increasingly unnecessary to wear crash helmets for what was more or less a Sunday drive in the country.

The car was in training to compete in the Pentagon's second Grand Challenge, an ambitious autonomous vehicle contest intended to jump-start technology planned for future robotic military vehicles. At the beginning of the twenty-first century, Congress instructed the U.S. military to begin designing autonomous vehicles. Congress even gave the Pentagon a specific goal: by 2015, one-third of the army's vehicles were supposed to go places without human drivers present. The directive wasn't clear as to whether both autonomous and remotely teleoperated vehicles would satisfy the requirement. In either case the idea was that smart vehicles would save both money and soldiers' lives. But by 2004, little progress had been made, and Tony Tether, the then controversial director of the Pentagon's blue-sky research arm, DARPA, the Defense Advanced Research Projects Agency, came up with a high-profile contest as a gambit to persuade computer hackers, college professors, and publicity-seeking corporations to innovate where the military had failed. Tether was a product of the military-industrial complex, and the contest itself was a daring admission that the defense contracting world was not able to get the job done. By opening the door for ragtag teams of hobbyists, Tether ran the risk of undermining the classified world dominated by the Beltway Bandits that surround Washington, D.C., and garner the lion's share of military research dollars.

The first Grand Challenge contest, held in 2004, was something of a fiasco. Vehicles tipped over, drove in circles, and ignominiously knocked down fences. Even the most successful entrant had gotten stuck in the dust just seven miles from the starting line in a 120-mile race, with one wheel spinning helplessly as it teetered off the edge of the road. When the dust settled, a reporter flying overhead in a light plane saw brightly colored vehicles scattered motionless over the desert floor. At the time it seemed obvious that self-driving cars were still years away, and Tether was criticized for organizing a publicity stunt.

Now, just a little more than a year later, Thrun was behind the wheel in a second-generation robot contestant. It felt like the future had arrived sooner than expected. It took only a dozen miles, however, to realize that techno-enthusiasm is frequently premature. Stanley crested a rise in the desert and plunged smartly into a swale. Then, as the car tilted upward, its laser guidance system swept across an overhanging tree limb. Without warning the robot navigator spooked, the car wrenched violently first left, then right, and instantly plunged off the road. It all happened faster than Thrun could reach over and pound the large red E-Stop button.

Luckily, the car found a relatively soft landing. The Touareg had been caught by an immense desert thornbush just off the road. It cushioned the crash landing and the car stopped slowly enough that the air bags didn't deploy. When the occupants surveyed the road from the crash scene, it was obvious that it could have been much worse. Two imposing piles of boulders bracketed the bush, but the VW had missed them.

The passengers stumbled out and Thrun scrambled up on top of the vehicle to reposition the sensors bent out of alignment by the crash. Then everyone piled back into Stanley, and Montemerlo removed an offending block of software code that had been intended to make the ride more comfortable for human passengers. Thrun restarted the autopilot and the machine once again headed out into the Arizona desert. There were other mishaps that day, too. The AI controller had no notion of the consequence of mud puddles and later in the day Stanley found itself ensnared in a small lake in the middle of the road. Fortunately there were several human-driven support vehicles nearby, and when the car's wheels began spinning helplessly, the support team of human helpers piled out to push the car out of the goo.

These were small setbacks for Thrun's team, a group of Stanford University professors, VW engineers, and student hackers among more than a dozen teams competing for a multimillion-dollar cash prize. The day was a low point after which things improved dramatically. Indeed, the DARPA contest would later prove to be a dividing line between a world in which robots were viewed as toys or research curiosities and one in which people began to accept that robots could move about freely.

Stanley's test drive was a harbinger of technology to come. The arrival of machine intelligence had been forecast for decades in the writings of science-fiction writers, so much so that when the technology actually began to appear, it seemed anticlimactic. In the late 1980s, anyone wandering through the cavernous Grand Central Station in Manhattan would have noticed that almost a third of the morning commuters were wearing Sony Walkman headsets. Today, of course, the Walkmans have been replaced by Apple's iconic bright white iPhone headphones, and there are some who believe that technology haute couture will inevitably lead to a future version of Google Glass—the search engine maker's first effort to augment reality—or perhaps more ambitious and immersive systems. Like the frog in the pot, we have been desensitized to the changes wrought by the rapid increase and proliferation of information technology.

The Walkman, the iPhone, and Google Glass all prefigure a world where the line between what is human and who is machine begins to blur. William Gibson's *Neuromancer,* the science-fiction novel that popularized the idea of cyberspace, drew a portrait of a new cybernetic territory composed of computers and networks. It also painted a future in which computers were not discrete boxes, but would be woven together into a dense fabric that was increasingly wrapped around human beings, "augmenting" their senses.

It is not such a big leap to move from the early-morning commuters wearing Sony Walkman headsets, past the iPhone users wrapped in their personal sound bubbles, directly to Google Glass–wearing urban hipsters watching tiny displays that annotate the world around them. They aren't yet "jacked into the net," as Gibson foresaw, but it is easy to assume that computing and communication technology is moving rapidly in that direction.

Gibson was early to offer a science-fiction vision of what has been called "intelligence augmentation." He imagined computerized inserts he called "microsofts"—with a lowercase *m*—that could be snapped into the base of the human skull to instantly add a particular skill—like a new language. At the time—several decades ago—it was obviously an impossible bit of science fiction. Today his cyborg vision is something less of a wild leap.

In 2013 President Obama unveiled the BRAIN initiative, an effort to simultaneously record the activities of one million neurons in the human brain. But one of the major funders of the BRAIN initiative is DARPA, and the agency is not interested in just reading *from* the brain. BRAIN scientists will patiently explain that one of the goals of the plan is to build a *two-way* interface between the human brain and computers. On its face, such an idea seems impossibly sinister, conjuring up images of the ultimate Big Brother and thought control. At the same time there is a utopian implication inherent in the technology. The potential future is perhaps the inevitable trajectory of human-computer interaction design, implicit in J. C. R. Licklider's 1960 manifesto, "Man-Computer Symbiosis," where he foretold a more intimate collaboration between humans and machines.

While the world of *Neuromancer* was wonderful science fiction, actually entering the world that Gibson portrayed presents a puzzle. On one hand, the arrival of cyborgs poses the question of what it means to be human. By itself that isn't a new challenge. While technology may be evolving increasingly rapidly today, humans have always been transformed by technology, as far back as the domestication of fire or the invention of the wheel (or its eventual application to luggage in the twentieth century). Since the beginning of the industrial era machines have displaced human labor. Now with the arrival of computing and computer networks, for the first time machines are displacing "intellectual" labor. The invention of the computer generated an earlier debate over the consequences of intelligent machines. The new wave of artificial intelligence technologies has now revived that debate with a vengeance.

Mainstream economists have maintained that over time the size of the workforce has continued to grow despite the changing nature of work driven by technology and innovation. In the nineteenth century, more than half of all workers were engaged in agricultural labor; today that number has fallen to around 2 percent—and yet there are more people working than ever in occupations outside of agriculture. Indeed, even with two recessions, between 1990 and 2010 the overall workforce in the United States increased by 21 percent. If the mainstream economists are correct, there is no economic cataclysm on a societal level due to automation in the offing.

However, today we are entering an era where humans can, with growing ease, be designed in or out of "the loop," even in formerly high-status, high-income, white-collar professional areas. On one end of the spectrum smart robots can load and unload trucks. On the other end, software "robots" are replacing call center workers and office clerks, as well as transforming high-skill, high-status professions such as radiology. In the future, how will the line be drawn between man and machine, and who will draw it?

Despite the growing debate over the consequences of the next generation of automation, there has been very little discussion about the designers and their values. When pressed, the computer scientists, roboticists, and technologists offer conflicting views. Some want to replace humans with machines; some are resigned to the inevitability—"I for one, welcome our insect overlords" (later "robot overlords") was a meme that was popularized by *The Simpsons*—and some of them just as passionately want to build machines to extend the reach of humans. The question of whether true artificial intelligence—the concept known as "Strong AI" or Artificial General Intelligence—will emerge, and whether machines can do more than mimic humans, has also been debated for decades. Today there is a growing chorus of scientists and technologists raising new alarms about the possibility of the emergence of self-aware machines and their consequences. Discussions about the state of AI technology today veer into the realm of science fiction or perhaps religion. However, the reality of machine autonomy is no longer merely a philosophical or hypothetical question. We have reached the point where machines are capable of performing many human tasks that require intelligence as well as muscle: they can do factory work, drive vehicles, diagnose illnesses, and understand documents, and they can certainly control weapons and kill with deadly accuracy.

The AI versus IA dichotomy is nowhere clearer than in a new generation of weapons systems now on the horizon. Developers at DARPA are about to cross a new technological threshold with a replacement for today's cruise missiles, the Long Range Anti-Ship Missile, or LRASM. Developed for the navy, it is scheduled for the U.S. fleet in 2018. Unlike its predecessors, this is a new weapon in the U.S. arsenal with the ability to make targeting decisions autonomously. The LRASM is designed to fly to an enemy fleet while out of contact with human controllers and then use artificial intelligence technologies to decide which target to kill.

The new ethical dilemma is, will humans allow their weapons to pull triggers on their own without human oversight? Variations of that same challenge are inherent in rapid computerization of the automobile, and indeed transportation in general is emblematic of the consequences of the new wave of smart machines. Artificial intelligence is poised to have an impact on society that will be greater than the effect that personal computing and the Internet have had beginning in the 1990s. Significantly, the transformation is being shepherded by a group of elite technologists.

Several years ago Jerry Kaplan, a Silicon Valley veteran who began his career as a Stanford artificial intelligence researcher and then became one of those who walked away from the field during the 1980s, warned a group of Stanford computer scientists and graduate student researchers: "Your actions today, right here in the Artificial Intelligence Lab, as embodied in the systems you create, may determine how society deals with this issue." The imminent arrival of the next generation of AI is a crucial ethical challenge, he contended: "We're in danger of incubating robotic life at the expense of our own life."[1] The dichotomy that he sketched out for the researchers is the gap between intelligent machines that displace humans and human-centered computing systems that extend human capabilities.

Like many technologists in Silicon Valley, Kaplan believes we are on the brink of the creation of an entire economy that runs largely without human intervention. That may sound apocalyptic, but the future Kaplan described will almost certainly arrive. His deeper point was that today's technology acceleration isn't arriving blindly. The engineers who are designing our future are each—individually—making choices.

On an abandoned military base in the California desert during the fall of 2007 a short, heavyset man holding a checkered flag stepped out onto a dusty makeshift racing track and waved it energetically as a Chevrolet Tahoe SUV glided past at a leisurely pace. The flag waver was Tony Tether, the director of DARPA.

There was no driver behind the wheel of the vehicle, which sported a large GM decal. Closer examination revealed no passengers in the car, and none of the other cars in the "race" had drivers or passengers either. Viewing the event, in which the cars glided seemingly endlessly through a makeshift town previously used for training military troops in urban combat, it didn't seem to be a race at all. It felt more like an afternoon of stop-and-go Sunday traffic in a science-fiction movie like *Blade Runner.*

Indeed, by almost any standard it was an odd event. The DARPA Urban Challenge pitted teams of roboticists, artificial intelligence researchers, students, automotive engineers, and software hackers against each other in an effort to design and build robot vehicles capable of driving autonomously in an urban traffic setting. The event was the third in the series of contests that Tether organized. At the time military technology largely amplified a soldier's killing power rather than replacing the soldier. Robotic military planes were flown by humans and, in some cases, by extraordinarily large groups of soldiers. A report by the Defense Science Board in 2012 noted that for many military operations it might take a team of several hundred personnel to fly a single drone mission.[2]

Unmanned ground vehicles were a more complicated challenge. The problem in the case of ground vehicles was, as one DARPA manager would put it, that "the ground was hard"— "hard" as in "hard to drive on," rather than as in "rock." Following a road is challenging enough, but robot car designers are confronted with an endless array of special cases: driving at night, driving into the sun, driving in rain, on ice—the list goes on indefinitely.

Consider the problem of designing a machine that knows

how to react to something as simple as a plastic bag in a lane on the highway. Is the bag hard, or is it soft? Will it damage the vehicle? In a war zone, it might be an improvised explosive device. Humans can see and react to such challenges seemingly without effort, when driving at low speed with good visibility. For AI researchers, however, solving that problem is the holy grail in computer vision. It became one of a myriad of similar challenges that DARPA set out to solve in creating the autonomous vehicle Grand Challenge events. In the 1980s roboticists in both Germany and the United States had made scattered progress toward autonomous driving, but the reality was that it was easier to build a robot to go to the moon than to build one that could drive by itself in rush-hour traffic. And so Tony Tether took up the challenge. The endeavor was risky: if the contests failed to produce results, the series of Grand Challenge self-driving contests would become known as Tether's Folly. Thus the checkered flag at the final race proved to be as much a victory lap for Tether as for the cars.

There had been darker times. Under Tether's directorship the agency hired Admiral John Poindexter to build the system known as Total Information Awareness. A vast data-mining project that was intended to hunt terrorists online by collecting and connecting the dots in oceans of credit card, email, and phone records, the project started a privacy firestorm and was soon canceled by Congress in May of 2003. Although Total Information Awareness vanished from public view, it in fact moved into the nation's intelligence bureaucracy only to become visible again in 2013 when Edward Snowden leaked hundreds of thousands of documents that revealed a deep and broad range of systems for surveillance of any possible activity that could be of interest. In the pantheon of DARPA directors, Tether was also something of an odd duck. He survived the Total Information Awareness scandal and pushed the agency ahead in other areas with a deep and controlling involvement in all of the agency's research projects. (Indeed, the decision by

Tether to wave the checkered flag was emblematic of his tenure at DARPA—Tony Tether was a micromanager.)

DARPA was founded in response to the Soviet Sputnik, which was like a thunderbolt to an America that believed in its technological supremacy. With the explicit mission of ensuring the United States was never again technologically superseded by another power, the directors of DARPA—at birth more simply named the Advanced Research Projects Agency—had been scientists and engineers willing to place huge bets on blue-sky technologies, with close relationships and a real sense of affection for the nation's best university researchers.

Not so with Tony Tether, who represented the George W. Bush era. He had worked for decades as a program manager for secretive military contractors and, like many surrounding George W. Bush, was wary of the nation's academic institutions, which he thought were too independent to be trusted with the new mission. Small wonder. Tether's worldview had been formed when he was an electrical engineering grad student at Stanford University during the 1960s, where there was a sharp division between the antiwar students and the scientists and engineers helping the Vietnam War effort by designing advanced weapons.

After arriving as director he went to work changing the culture of the agency that had gained a legendary reputation for the way it helped invent everything from the Internet to stealth fighter technology. He rapidly moved money away from the universities and toward classified work done by military contractors supporting the twin wars in Iraq and Afghanistan. The agency moved away from "blue sky" toward "deliverables." Publicly Tether made the case that it was still possible to innovate in secret, as long as you fostered the competitive culture of Silicon Valley, with its turmoil of new ideas and rewards for good tries even if they failed.

And Tether certainly took DARPA in new technology directions. His concern for the thousands of maimed veterans coming back without limbs and with increasing the power and effectiveness of military decision-makers inspired him to push agency dollars into human augmentation projects as well as artificial intelligence. That meant robotic arms and legs for wounded soldiers, and an "admiral's advisor," a military version of what Doug Engelbart had set out to do in the 1960s with his vision of intelligence augmentation, or IA. The project was referred to as PAL, for Perceptive Assistant that Learns, and much of the research would be done at SRI International, which dubbed the project CALO, or Cognitive Assistant that Learns and Organizes.

It was ironic that Tether returned to the research agenda originally promoted during the mid-1960s by two visionary DARPA program managers, Robert Taylor and J. C. R. Licklider. It was also bittersweet, although few mentioned it, that despite Doug Engelbart's tremendous early success in the early 1970s, his project had faltered and fallen out of favor at SRI. He ended up being shuffled off to a time-sharing company for commercialization, where his project sat relatively unnoticed and underfunded for more than a decade. The renewed DARPA investment would touch off a wave of commercial innovation—CALO would lead most significantly to Apple's Siri personal assistant, a direct descendant of the augmentation approach originally pioneered by Engelbart.

Tether's automotive Grand Challenge drew garage innovators and eager volunteers out of the woodwork. In military terms it was a "force multiplier," allowing the agency to get many times the innovation it would get from traditional contracting efforts. At its heart, however, the specific challenge that Tether chose to pursue had been cooked up more than a decade earlier inside the same university research community that he now disfavored. The guiding force behind the GM robot SUV that would win the Urban Challenge in 2007 was a Carnegie Mellon roboticist who had been itching to win this prize for more than a decade.

In the fall of 2005, Tether's second robot race through the California desert had just ended at the Nevada border and Stanford University's roboticists were celebrating. Stanley, the once crash-prone computerized Volkswagen Touareg, had just pulled off a come-from-behind victory and rolled under a large banner before a cheering audience of several thousand.

Just a few feet away in another tent, however, the atmosphere had the grim quality of a losing football team's locker room. The Carnegie Mellon team had been the odds-on favorite, with two robot vehicle entries and a no-nonsense leader, a former marine and rock climber, William L. "Red" Whittaker. His team had lost the race due to a damnable spell of bad luck. Whittaker had barnstormed into the first DARPA race eighteen months earlier with another heavily funded GM Humvee, only to fail when the car placed a wheel just slightly off road on a steep climb. Trapped in the sand, it was out of the competition. Up to then, Whittaker's robot had been head and shoulders above the others. So when he returned the second time with a two-car fleet and a squad of photo analysts to pore over the course ahead of the competition, he had easily been cast as the odds-on favorite.

Once again, however, bad luck struck. His primary vehicle led Stanley until late in the race, when it malfunctioned, slowing dramatically and allowing the Stanford team to sail by and grab the $2 million prize. After the second loss, Whittaker stood in the tent in front of his team and gave an inspiring speech worthy of any college football coach. "On any Sunday . . ." he told his team, echoing the words of the losing coach. The loss was especially painful because the leaders of the Stanford team, Sebastian Thrun and Mike Montemerlo, were former CMU roboticists who had defected to Stanford, where they organized the rival, winning effort. Years later the loss still rankled. Outside of Whittaker's office at the university is a portrait of the ill-fated team of robot car designers. In the hallway Whittaker would greet visitors and replay the failure in detail.

The defeat was particularly striking because Red Whittaker had in many ways been viewed widely as the nation's premier roboticist. By the time of the Grand Challenges he had already become a legend for designing robots capable of going places where humans couldn't go. For decades he combined a can-do attitude with an adventurer's spirit. His parents had both flown planes with a bit of barnstorming style. His father, an air force bomber pilot, sold mining explosives after the war. His mother, a chemist, was a pilot, too. When he was a young man, she had once flown him under a bridge.[3]

His Pennsylvania upbringing led him to develop a style of robotics that pushed in the direction of using the machines primarily as tools to extend an adventurer's reach, a style in the tradition of Yvon Chouinard, the legendary climber who designed and made his own climbing hardware, or Jacques Cousteau, the undersea explorer who made his own breathing equipment. With a degree in civil engineering from Princeton and a two-year tour as a marine sergeant, the six-foot-four Whittaker pioneered "field" robotics—building machines that left the laboratory and moved around in the world.

In Red Whittaker's robotic world, however, humans were still very much in the loop. In every case he used them to extend his reach as an adventurer. He had built machines used in nuclear power plant catastrophes at both Three Mile Island and Chernobyl. In the late 1980s he designed a huge nineteen-foot-tall robot called Ambler that was intended to walk on Mars. He sent a robot into a volcano and had been one of the first roboticists in the United States to explore the idea of an autonomous car as part of Carnegie Mellon's Navlab project.

"This is not the factory of the future," he was fond of pointing out. "The ideas that make it in the factory don't make it in the outside world."[4]

As a young man Whittaker had variously been a rower, wrestler, boxer, and mountain climber. His love of adventure had not been without personal pain, however. He spent a decade of his life rock climbing, sneaking away from his robot projects to spend time in Yosemite and the Himalayas. He even soloed the east wall of the Matterhorn in winter conditions. He had begun climbing casually as a member of a local explorer's club in Pittsburgh. It only became a passion when he met another young climber, after seeing a notice on a bulletin board: "Expert climber willing to teach the right guy," the note read, adding: "You must have a car."

The two would become inseparable climbing partners over the next decade.

That magic time for Whittaker came to a sudden end one summer when they were climbing in Peru. His Pittsburgh friend was climbing with another young climber. The two were roped together and the younger man slipped and pulled both men down a tumbling set of ledges for almost a thousand feet. Whittaker, who was off-rope during the accident, was able to rescue the young climber, but his friend was killed by the fall. Whittaker returned to Pittsburgh shaken by the accident. It would take months before he mustered up the courage to go over to the home where the young man had lived with his parents and clean out the dead climber's room.

The death left its mark. Whittaker stopped climbing, but still hungered for some sort of challenging adventure. He began to build ever more exotic robots, capable of performing tasks ranging from simple exploration to sophisticated repair, to extend his adventures into volcanoes, and ultimately, perhaps, to the moon and Mars. Even when he had been climbing on Earth in the 1970s and 1980s, it was becoming more and more difficult to find virgin territory. With the possibility of "virtual exploration," new vistas would open up indefinitely and Whittaker could again dream of climbing and rappelling, this time perhaps with a humanoid robot stand-in on another world.

Whittaker redeemed his bitter loss to Stanford's Stanley several years later in the third Grand Challenge, in 2007. His General Motors–backed "Boss" would win the final Urban Driving Challenge.

One of the most enduring bits of Silicon Valley lore recalls how Steve Jobs recruited Pepsi CEO John Sculley to Apple by asking him if he wanted to spend the rest of his life selling sugar water. Though some might consider it naive, the Valley's ethos is about changing the world. That is at the heart of the concept of "scale," which is very much a common denominator in motivating the region's programmers, hardware hackers, and venture capitalists. It is not enough to make a profit, or to create something that is beautiful. It has to have an impact. It has to be something that goes under 95 percent of the world's Christmas trees, or offers clean water or electricity to billions of people.

Google's chief executive Larry Page took the Steve Jobs approach in recruiting Sebastian Thrun. Thrun was a fast-rising academic who had spent a sabbatical year at Stanford in 2001, which opened his eyes to the world that Silicon Valley offered beyond the walls of academia. There was more out there besides achieving tenure, publishing, and teaching students.

He returned to Stanford as an assistant professor in 2003. He attended the first DARPA Grand Challenge as an observer. The self-driving car competition completely changed his perspective: he realized that there were great thinkers outside of his cloistered academic community who cared deeply about changing the world. In between, during his short return to CMU, he had sent a note to Whittaker offering to help their software effort, but was rebuffed. Thrun had brought a group of students with him from CMU when he returned to Stanford, including Mike Montemerlo, whose father was a NASA roboticist. Montemerlo gave a presentation on the first DARPA contest.

Sebastian Thrun (*left*) and Mike Montemerlo (*right*) in front of the Stanford University autonomous vehicle while it was being tested to take part in DARPA's Urban Challenge in 2007. (*Photo courtesy of the author*)

At the end of his presentation his final slide asked, "Should we at Stanford enter the Grand Challenge?" And then he answered his own question in a large font. "NO!" There were a dozen reasons not to do it. They would have no chance of winning, it was too hard, it would cost too much money. Thrun looked at Montemerlo and it was obvious that although on paper he was the quintessential pessimist, everything in his demeanor was saying yes.

Soon afterward Thrun threw himself into the DARPA competition with passion. For the first time in his life he felt like he was focusing on something that was genuinely likely to have broad impact. Living in the Arizona desert for weeks on end, surviving on pizza, the team worked on the car until it was able to drive the backcountry roads flawlessly.

Montemerlo and Thrun made a perfect team of opposites. Montemerlo was fundamentally conservative, and Thrun was extraordinarily risk-inclined. As head of software, Montemerlo would build his conservative assumptions into his programs. When he wasn't looking, Thrun would go through the code and comment out the limitations to make the car go faster. It would infuriate the younger researcher. But in the end it was a winning combination.

Larry Page had said to Thrun that if you really focus on something you can achieve amazing things. He was right. After Stanley captured the $2 million DARPA prize, Thrun took Page's words to heart. The two men had become friends after Thrun helped the Google cofounder debug a home robot that Page had been tinkering with. Thrun borrowed the device and brought it back able to navigate inside Page's home.

Navigation, a necessity for autonomous robots, had become Thrun's particular expertise. At CMU and later at Stanford he worked to develop SLAM, the mapping technique pioneered at Stanford Research Institute by the designers of the first mobile robots beginning in the 1960s. Thrun had helped make the technique fast and accurate and had paved the way for using it in autonomous cars. At Carnegie Mellon he had begun to attract national attention for a variety of mobile robots. In 1998 at the Smithsonian in D.C., he showcased Minerva, a mobile museum tour guide that was connected to the Web and could interact with museum guests and travel up to three and a half miles per hour. He worked with Red Whittaker to send robots into mines, which relied heavily on SLAM techniques. Thrun also tried to integrate mobile and autonomous robots in nursing and elder-care settings, with little success. It turned out to be a humbling experience, which gave him a deep appreciation of the limitations of using technologies to solve human problems. In 2002, in a team effort between the two universities, Thrun pioneered a new flavor of SLAM that was dubbed Fast-SLAM, which could be used in real-world situations where it was necessary to locate thousands of objects. It was an early example of a new wave of artificial intelligence and robotics that increasingly relied on probabilistic statistical techniques rather than on rule-based inference.

At Stanford, Thrun would rise quickly to become director of

the revitalized Stanford Artificial Intelligence Laboratory that had originally been created by John McCarthy in the 1960s. But he also quickly became frustrated by the fragmented life of an academic, dividing time between teaching, public speaking, grant writing, working on committees, doing research, and mentoring. In the wake of his 2005 DARPA Grand Challenge victory Thrun had also become more visible in high-technology circles. His talks described the mass atrocities committed by human drivers that resulted in more than one million killed and maimed each year globally. He personalized the story. A close friend had been killed in an automobile accident when Thrun was a high school student in his native Germany. Many people he was close to lost friends in accidents. More recently, a family member of a Stanford faculty secretary was crippled for life after a truck hit her car. In an instant she went from being a young girl full of life and possibility to someone whose life was forever impaired. Thrun's change-the-world goals gave him a platform at places like the TED Conference.

After building two vehicles for the DARPA Challenge contests, he decided to leave Stanford. Page offered him the opportunity to do things at "Google scale," which meant that his work would touch the entire world. He secretly set up a laboratory modeled vaguely on Xerox PARC, the legendary computer science laboratory that was the birthplace of the modern personal computer, early computer networks, and the laser printer, creating projects in autonomous cars and reinventing mobile computing. Among other projects, he helped launch Google Glass, which was an effort to build computing capabilities including vision and speech into ordinary glasses.

Unlike laboratories of the previous era that emphasized basic science, such as IBM Research and Bell Labs, Google's X Lab was closer in style to PARC, which had been established to vault the copier giant, restyled "the Document Company," into the computer industry—to compete directly with IBM. The X Lab was intended to push Google into new markets. Google felt secure in its Web search monopoly so, with a profit stream that by the end of 2013 was more than $1 billion a month, the search company funded ambitious R & D projects that might have nothing to do with the company's core business. Google was famous for its 70-20-10 rule, which gave its engineers free time to pursue their own side projects. Employees are supposed to spend 10 percent of their time on projects entirely unrelated to the company's core business. Its founders Sergey Brin and Larry Page believed deeply in thinking big. They called their efforts "moon shots": not pure science, but research projects that were hopefully destined to have commercial rather than purely scientific impact.

It was a perfect environment for Thrun. His first project in 2008 had been to create the company's fleet of Street View cars that systematically captured digital images of homes and businesses on every street in the nation. The next year he began an even more ambitious effort: a self-driving car that would travel on public streets and highways. He was both cautious and bold in the car project. A single accident might destroy the Google car, so at the outset he ensured that a detailed safety regime was in place. He was acutely aware that if there was any indication in the program that Google had not been incredibly careful, it would be a disaster. He never let an untrained driver near the wheel of the small Toyota Prius fleet on which the system was being developed. The cars would eventually drive more than a half-million miles without an accident, but Thrun understood that even a single error every fifty thousand to a hundred thousand miles was too high an error rate. At the same time he believed that there was a path forward that would allow Google to redefine what it meant to be in a car.

Like the automotive industry, Thrun and his team believed in the price/volume curve, which suggested that costs would go down the more a company manufactured a particular thing. Sure, today a single experimental lidar laser radar might cost tens of thousands of dollars, but the Google engineers had faith that in a few years it would be so cheap that it would not be a showstopper in the bill of materials of some future car. In the trade-off between cost and durability, Thrun always felt it would make sense to design and build more reliable systems now and depend on mass manufacturing technologies for price reductions to kick in later. The pricey laser guidance systems didn't actually contain that many parts, so there was little reason to believe that prices couldn't come down rapidly. It had already happened with radar, which had once been an esoteric military and aviation technology but in recent years had begun showing up in motion detectors and premium automobiles.

Thrun evinced an engineer's worldview and tended toward a libertarian outlook. He held a pro-business point of view that the global corporation was an evolutionary step beyond the nation-state. He also subscribed to the belief, commonplace in the Valley, that within three decades as much as 90 percent of all jobs will be made obsolete by advancing AI and robotic technologies. Indeed, Thrun believed that most people's jobs are actually pretty useless and unfulfilling. There are countless manual labor jobs—everything from loading and unloading trucks to driving them—that could vanish over the coming decade. He also believed that much of the bureaucratic labor force is actively counterproductive. Those people make other people's work harder. Thrun had a similar contempt for what he perceived as Detroit's hidebound car industry that could have easily used technology to radically reshape transportation systems and make them safer, but did little and was content to focus on changing the shape of a car's tail fins each year. By 2010 he had a deep surprise in store for an industry that did not change easily and was largely unfamiliar with Silicon Valley culture.[5]

The DARPA races created ripples in Detroit, the cradle of the American automotive industry, but the industry kept to its traditional position that cars were meant to be driven by people and should not operate autonomously. By and large the industry had generally resisted computer technology. Many car manufacturers adhered to a "computers are buggy" philosophy. However, engineers elsewhere in the country were beginning to think about transportation through the new lens of cheap sensors, the microprocessor, and the Internet.

In the spring of 2010, rumors about an experimental Google car began to float around Silicon Valley. Initially they sounded preposterous. The company, nominally a provider of Internet search, was supposedly hiding the cars in plain sight. Google engineers, so the story went, had succeeded in robotically driving from San Francisco to Los Angeles on freeways at night! The notion immediately elicited both guffaws and pointed reminders that such an invention would be illegal, even if it was possible. How could they get away with something so crazy?

Of course, Google's young cofounders Sergey Brin and Larry Page had by then perfected a public image for wild schemes based on AI and other futuristic technologies to transform the world. Eric Schmidt, the company's chief executive officer beginning in 2001, would tell reporters that his role was one of adult supervision—persuading the cofounders which of their ideas should be kept above and which below the "bar." The cofounders famously considered the idea of a space elevator. New, incredibly strong material had recently been developed, and this material was so strong that, rather than using a rocket, it would be possible to build a cable that reached from the Earth into orbit to inexpensively hoist people and materials into space. When queried about the idea Schmidt would pointedly state that this was one of the ideas that was being considered, but was—for the moment at least—"below the bar."

In the hothouse community of technical workers that is Silicon Valley, however, it is difficult to keep secrets. It was obvious that something was afoot. Within a year after the final DARPA Grand Challenge event in 2007, Sebastian Thrun had taken a leave from Stanford and gone to work full-time at Google. His departure was never publicly announced, or even mentioned in the press, but among the Valley's digerati, Thrun's change of venue was noted with intense interest. A year later, while he was with colleagues in a bar in Alaska at an artificial intelligence conference, he spilled out a few tantalizing comments. Those words circulated back in Silicon Valley and made people wonder.

In the end, however, it was a high school friend of one of the low-paid drivers the company had hired to babysit its robotic Prius fleet who inadvertently spilled the beans. *One of the kids I went to high school with is being paid fifteen dollars an hour by Google to sit in a car while it drives itself!* a young college student blurted to me. At that point the secret became impossible to contain. The company was parking its self-driving cars in the open lots on the Google campus.

The Google engineers had made no effort to conceal the sensors attached to the roof of the ungainly-looking creatures, which looked even odder than their predecessor, Stanford's Stanley. Rather than an array of sensors mounted above the windshield, each Prius had a single 360-degree lidar, mounted a foot above the center of the car's roof. The coffee-can-sized mechanical laser, made by Velodyne, a local high-tech company, made it possible to easily create a real-time map of the surrounding environment for several hundred feet in all directions. It wasn't cheap—at the time the lidar alone added $70,000 to the vehicle cost.

How did the odd-looking Toyotas, also equipped with less obtrusive radars, cameras, GPS, and inertial guidance sensors, escape discovery for as long as they did? There were several reasons. The cars were frequently driven at night, and the people who saw them confused them with a ubiquitous fleet of Google Street View cars, which had a large camera on a mast above the roof taking photographs that were used to build a visual map of the surrounding street as the car drove. (They also recorded people's Wi-Fi network locations, which then could be used as beacons to improve the precision in locating Google's Android smartphones.)

The Street View assumption usually hid the cars in plain sight, but not always. The Google engineer who had the pleasure of the first encounter with law enforcement was James Kuffner, a former CMU roboticist who had been one of the first members of the team. Kuffner had made a name for himself at Carnegie Mellon working on both navigation and a variety of humanoid robot projects. His expertise was in motion planning, figuring out how to teach machines to navigate in the real world. He was bitten by the robot car bug as part of Red Whittaker's DARPA Grand Challenge team, and when key members of that group began to disappear into a secret Google project code-named Chauffeur, he jumped at the chance.

Late one night they were testing the robotic Prius in Carmel, one of the not-quite-urban driving areas they were focusing on closely. They were testing the system late at night because they were anxious to build detailed maps with centimeter accuracy, and it was easier to get baseline maps of the streets when no one was around. After passing through town several times with their distinctive lidar prominently displayed, Kuffner was sitting in the driver's seat when the Prius was pulled over by a local policeman suspicious about the robot's repeated passes.

"What is this?" he asked, pointing to the roof.

Kuffner, like all of the Google drivers, had been given strict instructions how to respond to this inevitable confrontation. He reached behind him and handed a prewritten document to the officer. The police officer's eyes widened as he read it. Then he grew increasingly excited and kept the Google engineers chatting late into the night about the future of transportation.

The incident did not lead to public disclosure, but once I discovered the cars in the company's parking lots while reporting for the *New York Times,* the Google car engineers relented and offered me a ride.

From a backseat vantage point it was immediately clear that in the space of just three years, Google had made a significant leap past the cars of the Grand Challenge. The Google Prius replicated much of the original DARPA technology, but with more polish. Engaging the autopilot made a whooshing *Star Trek* sound. Technically, the ride was a remarkable tour de force. A test drive began with the car casually gliding away from Google's campus on Mountain View city streets. Within a few blocks, the car had stopped at both stop signs and stoplights and then merged onto rush-hour traffic on the 101 freeway. At the next exit the car then drove itself off the freeway onto a flyover overpass that curved gracefully over the 101. What was most striking to the first-time passenger was the car's ability to steer around the curve exactly as a human being might. There was absolutely nothing robotic about AI's driving behavior.

When the *New York Times* published the story, the Google car struck Detroit like a thunderbolt. The automobile industry had been adding computer technology and sensors to cars at a maddeningly slow pace. Even though cruise control had been standard for decades, intelligent cruise control—using sensors to keep pace with traffic automatically—was still basically an exotic feature in 2010. A number of automobile manufacturers had outposts in Silicon Valley, but in the wake of the publicity surrounding the Google car, the remaining carmakers rushed to build labs close by. Nobody wanted to see a repeat of what happened to personal computer hardware makers when Microsoft Windows became an industry standard and hardware manufacturers found that their products were increasingly low-margin commodities while much of the profit in the industry flowed to Microsoft. The automotive industry now realized that it was facing the same threat.

At the same time, the popular reaction to the Google car was mixed. There had long been a rich science-fiction tradition of *Jetsons*-like futuristic robot cars. They had even been the stuff of TV series like *Knight Rider,* a 1980s show featuring a crime fighter assisted by an artificially intelligent car. There was also a dark-side vision of automated driving, perhaps best expressed in Daniel Suarez's 2009 sci-fi thriller *Daemon,* in which AI-controlled cars not only drove themselves, but ran people down as well. Still, the general perception was a deep well of skepticism about whether driverless cars would ever become a reality. However, Sebastian Thrun had made his point abundantly clear that humans are terrible drivers, largely the consequence of human fallibility and inattention. By the time his project was discovered, Google cars had driven more than a hundred thousand miles without an accident, and over the next several years that number would rise above a half-million miles. A young Google engineer, Anthony Levandowski, routinely commuted from Berkeley to Mountain View, a distance of fifty miles, in one of the Priuses, and Thrun himself would let a Google car drive him from Mountain View to his vacation home in Lake Tahoe on weekends.

Today, partially autonomous cars are already appearing on the market, and they offer two paths toward the future of transportation—one with smarter and safer human drivers and one in which humans will become passengers.

Google had not disclosed how it planned to commercialize its research, but by the end of 2013 more than a half-dozen automakers had already publicly stated their intent to offer autonomous vehicles. Indeed, 2014 was the year that the line was first crossed commercially when a handful of European car manufacturers including BMW, Mercedes, Volvo, and Audi announced an optional feature—traffic jam assist, the first baby step toward autonomous driving. In Audi's case, while on the highway, the car will drive autonomously when traffic is moving at less than forty miles per hour, staying in its lane and requiring driver intervention only as dictated by lawyers fearful that passengers might go to sleep or otherwise distract themselves. In late 2014 Tesla announced that it would begin to offer an "autopilot" system for its Model S, making the car self-driving in some highway situations.

The autonomous car will sharpen the dilemma raised by the AI versus IA dichotomy. While there is a growing debate over the liability issue—who will pay when the first human is killed by a robot car—the bar that the cars must pass to improve safety is actually incredibly low. In 2012 a National Highway Transportation Safety Administration study estimated that the deployment of electronic stability control (ESC) systems in light vehicles alone would save almost ten thousand lives and prevent almost a quarter million injuries.[6] Driving, it would seem, might be one area of life where humans should be taken out of the loop to the greatest degree possible. Even unimpaired humans are not particularly good drivers, and we are worse when distracted by the gadgets that increasingly surround us. We will be saved from ourselves by a generation of cheap cameras, radars, and lidars that, when coupled with pattern-sensing computers, will wrap an all-seeing eye around our cars, whether we are driving or are being driven.

For Amnon Shashua, the aha moment came while seated in a university library as a young computer science undergraduate in Jerusalem. Reading an article written in Hebrew by Shimon Ullman, who had been the first Ph.D. student under David Marr, a pioneer in vision research, he was thrilled to discover that the human retina was in many ways a computer. Ullman was a computer scientist who specialized in studying vision in both humans and machines. The realization that computing was going on inside the eye fascinated Shashua and he decided to follow in Ullman's footsteps.

He arrived at MIT in 1996 to study artificial intelligence when the field was still recovering from an earlier cycle of boom-and-bust. Companies had tried to build commercial expert systems based on the rules and logic approach of early artificial intelligence pioneers like Ed Feigenbaum and John McCarthy. In the heady early days of AI it had seemed that it would be straightforward to simply bottle the knowledge of a human expert, but the programs were fragile and failed in the marketplace, leading to the collapse of a number of ambitious start-ups. Now the AI world was rebounding. Progress in AI, which had been relatively stagnant for its first three decades, finally took off during the 1990s because statistical techniques made classification and decision-making practical. AI experiments hadn't yet seen great results because the computers of the era were still relatively underpowered for the data at hand. The new ideas, however, were in the air.

As a graduate student Shashua would focus on a promising approach to visually recognizing objects based on imaging them from multiple views to capture their geometry. The approach was derived from the world of computer graphics, where Martin Newell had pioneered a new modeling approach as a graduate student at the University of Utah—which was where much of computer graphics was invented during the 1970s. A real Melitta teapot found in his kitchen inspired Newell's approach. One day, as he was discussing the challenges of modeling objects with his wife over tea, she suggested that he model that teapot, which thereafter became an iconic image in the early days of computer graphics research.

At MIT, Shashua studied under computer vision scientists Tommy Poggio and Eric Grimson. Poggio was a scientist who stood between the worlds of computing and neuroscience and Grimson was a computer scientist who would later become MIT's chancellor. At the time there seemed to be a straight path from capturing shapes to recognizing them, but programming the recognition software would actually prove daunting. Even today the holy grail of "scene understanding"—for example, not only identifying a figure as a woman but also identifying what she might be doing—is still largely beyond reach, and significant progress has been made only in niche industries. For example, many cars can now identify pedestrians or bicyclists in time to automatically slow before a collision.

Shashua would become one of the masters in pragmatically carving out those niches. In an academic world where brain scientists debated computational scientists, he would ally himself with a group who took the position that "just because airplanes don't flap their wings, it doesn't mean they can't fly." After graduate school he moved back to Israel. He had already founded a successful company, Cognitens, using vision modeling to create incredibly accurate three-dimensional models of parts for industrial applications. The images, accurate to hair-thin tolerances, gave manufacturers ranging from automotive to aerospace the ability to create digital models of existing parts, enabling checking their fit and finish. The company was quickly sold.

Looking around for another project, Shashua heard from a former automotive industry customer about an automaker searching for stereovision technology for computer-assisted driving. They knew about Shashua's work in multiple-view geometry and asked if he had ideas for stereovision. He responded, "Well, that's fine but you don't need a stereo system, you can do it with a single camera." Humans can tell distances with one eye shut under some circumstances, he pointed out.

The entrepreneurial Shashua persuaded General Motors to invest $200,000 to develop demonstration software. He immediately called a businessman friend, Ziv Aviram, and proposed that they start a new company. "There is an opportunity," he told his friend. "This is going to be a huge field and everybody is thinking about it in the wrong way and we already have a customer, somebody who is willing to pay money." They called the new company Mobileye and Shashua wrote software for the demonstration on a desktop computer, soon showing one-camera machine vision that seemed like science fiction to the automakers at that time.

Six months after starting the project, Shashua heard from a large auto industry supplier that General Motors was about to offer a competitive bid for a way to warn drivers that the vehicle was straying out of its lane. Until then Mobileye had been focusing on far-out problems like vehicle and pedestrian detection that the industry thought weren't solvable. However, the parts supplier advised Shashua, "You should do something now. It's important to get some real estate inside the vehicle, then you can build more later."

The strategy made sense to Shashua, and so he put one of his Hebrew University students on the project for a couple of months. The lane-keeping software demonstration wasn't awful, but he realized it probably wasn't as good as what companies who'd started earlier could show, so there was virtually no way that the fledgling company would win.

Then he had a bright idea. He added vehicle detection to the software, but he told GM that the capability was a bug and that they shouldn't pay attention. "It will be taken out in the next version, so ignore it," he said. That was enough. GM was ecstatic about the safety advance that would be made possible by the ability to detect other vehicles at low cost. The automaker immediately canceled the bidding and committed to fund the novice firm's project developments. Vehicle detection would facilitate a new generation of safety features that didn't replace drivers, but rather augmented them with an invisible sensor and computer safety net. Technologies like lane departure warning, adaptive cruise control, forward collision warning, and anticollision braking are now rapidly moving toward becoming standard safety systems on cars.

Mobileye would grow into one of the largest international suppliers of AI vision technology for the automotive industry, but Shashua had bigger ideas. After creating Cognitens and Mobileye, he took a postdoctoral year at Stanford in 2001 and shared an office with Sebastian Thrun. Both men would eventually pioneer autonomous driving. Shashua would pursue the same technologies as Thrun, but with a more pragmatic, less "moon shot" approach. He had been deeply influenced by Poggio, who pursued biological approaches to vision, which were alternatives to using the brute force of increasingly powerful computers to recognize objects.

The statistical approach to computing would ultimately work best when both powerful clusters of computers, such as Google's cloud, and big data sets were available. But what if you didn't have those resources? This is where Shashua would excel. Mobileye had grown to become a uniquely Israeli technology firm, located in Jerusalem, close to Hebrew University, where Shashua teaches computer science. A Mobileye-equipped Audi served as a rolling research platform. Unlike the Google car, festooned with sensors, from the outside the Mobileye Audi looked normal, apart from a single video camera mounted unobtrusively just in front of the rearview mirror in the center of the windshield. The task at hand—automatic driving—required powerful computers, hidden in the car's trunk, with some room left over for luggage.

Like Google, Mobileye has significant ambitions that are still only partially realized. On a spring afternoon in 2013, two Mobileye engineers, Gaby Hayon and Eyal Bagon, drove me several miles east of Jerusalem on Highway 1 until they pulled off at a nondescript turnout where another employee waited in a shiny white Audi A7. As we got in the A7 and prepared for a test drive, Gaby and Eyal apologized to me. The car was a work in progress, they explained. Today Mobileye supplies computer vision technology to automakers like BMW, Volvo, Ford, and GM for safety applications. The company's third-generation technology is touted as being able to detect pedestrians and cyclists. Recently, Nissan gave a hint of things to come, demonstrating a car that automatically swerved to avoid a pedestrian walking out from behind a parked car.

Like Google, the Israelis are intent on going further, developing the technology necessary for autonomous driving. But while Google might decide to compete with the automobile industry by partnering with an upstart like Tesla, Shashua is exquisitely sensitive to the industry culture exemplified by its current customers. That means that his vision system designs must cost no more than several hundred dollars for even a premium vehicle and less than a hundred for a standard Chevy.

Google and Mobileye have taken very different approaches to solving the problem of making a car aware of its surroundings with better-than-human precision at highway speeds. Google's system is based on creating a remarkably detailed map of the world around the car using radars, video, and a Velodyne lidar, all at centimeter accuracy, augmenting the data it collects using its Street View vehicles. The Google car connects to the map database via a wireless connection to the Google cloud. The network is an electronic crutch for the car's navigation system, confirming what the local sensors are seeing around the car.

The global map database could make things easier for Google. One of the company's engineers confided that when the project got under way the Google team was surprised to find how dynamic the world is. Not only do freeway lanes frequently come and go for maintenance reasons, but "whole bridges will move," he said. Even without the database, the Google car is able to do things that might seem to be the province of humans alone, such as seamlessly merging into highway traffic and handling stop-and-go traffic in a dense urban downtown.

Google has conducted its project with a mix of Thrun's German precision and the firm's penchant for secrecy. The Israelis are more informal. On that hot spring afternoon in suburban Jerusalem there was little caution on the part of the Mobileye engineers. "Why don't you drive?" Eyal suggested to me, as he slid into the passenger seat behind a large display and keyboard. The engineers proceeded to give a rapid-fire minute-long lesson on driving a robot car: You simply turn on cruise control and then add the lane-keeping feature—steering—by pulling the cruise control stick on the steering wheel toward you. A heads-up display projected on the windshield showed the driver the car's speed and an icon indicated that the autonomous driving feature was on.

Unlike the Google car, which originally had a distinctive *Star Trek* start-up sound, there is only a small visual cue when the autopilot is engaged and the Mobileye Audi takes off down the highway by itself, at times reaching speeds of more than sixty miles per hour. On the road that snakes down a desolate canyon to the Dead Sea, it is difficult to relax. In an automated car, it is very challenging for a novice driver when, before long, a car ahead begins to slow for a stoplight. It takes all of one's willpower to keep a foot off the brake and trust the car as, sure enough, it slows down and rolls smoothly to a stop behind the vehicle ahead.

The Google car conveys the detached sense of a remote and slightly spooky machine intelligence at work somewhere in the background or perhaps in some distant cloud of computers. By contrast, during its test phase in 2013, the Mobileye car left a passenger acutely aware of the presence of machine assistance. The car needs to weave a bit within the lane when it starts to pull away from a stop—not a behavior that inspires confidence. If you understand the underlying technology, however, it ceases to be alarming. The Audi's vision system uses a single "monocular" camera. The third dimension, depth, is computed based on a clever algorithm that Shashua and his researchers designed, referred to as "structure from motion," and by weaving slightly the car is able to build a 3-D map of the world ahead of it.

Knowing that did little to comfort a first-time passenger, however. During the test ride, as it passed a parked car, the Audi pulled in the direction of the vehicle. Not wanting to see what the car was "really thinking," I grabbed the wheel and nudged the Audi back into the center of the lane. The Israeli engineers showed no signs of alarm, only amusement. After a half hour's drive along an old road that felt like it was still a part of antiquity, the trip was over. The autonomous ride felt jarringly like science fiction, yet it was just the first hint of what will gradually become a broad societal phase change. "Traffic jam assist" has already appeared on the market. Technology that was remarkable to someone visiting Israel in 2013 routinely handles stop-and-go freeway traffic around the world today.

The next phase of automatic driving will start arriving well before 2020—vehicles will handle routine highway driving, not just in traffic jams, but also in the commute from on-ramp to off-ramp. General Motors now calls this capability "Super Cruise," and it will mark a major step in the changing role of humans as drivers—from manual control to supervision.

The Google vision is clearly to build a vehicle in which the human becomes a passenger and no longer participates in driving. Yet Shashua believes that even for Google the completely driverless vehicle is still far in the future. Such a car will stumble across what he describes as the four-way stop puzzle—a completely human predicament. At intersections without stoplights there is an elaborate social dance that goes on between drivers, and it will be difficult for independent, noncommunicating computer systems to solve anytime in the foreseeable future.

Another complication is that human drivers bend the rules and ignore the protocols frequently, and pedestrians add massive complications. These challenges may be the real barrier for a future of completely AI-based automobiles in urban settings: we haven't yet been able to wrap our heads around the legal trouble posed by a possible AI-caused accident. There is a middle ground, Shashua believes, somewhere short of the Google vision, but realistic enough that it will begin to take over highway driving in just a couple of years. His approach wraps increasingly sophisticated sensor arrays and AI software around the human driver, who stays in the loop, a human with superawareness who can see farther and more clearly, and perhaps switch back and forth from other tasks besides driving. The car can alert the driver when his or her participation is beneficial or necessary, depending on the driver's preferences. Or perhaps the car's preferences.

Standing beside the Audi in the Jerusalem suburbs, it was clear that this was the new Promised Land. Like it or not, we are no longer in a biblical world, and the future is not about geographical territory, but rather about a rapidly approaching technological wonder world. Machines that will begin as golems are becoming brilliant and capable of performing many human tasks, from physical labor to rocket science.

Google had a problem. More than three years into the company's driverless car program, the small research team at the Mountain View–based Internet search company had safely driven more than a half-million miles autonomously. They had made startling progress in areas that had been generally believed impossible within the traditional automotive industry. Google cars could drive during the day and at night, could change lanes, and could even navigate the twisty Lombard Street in San Francisco. Google managed these advances by using the Internet to create a virtual infrastructure. Rather than building "smart" highways that would entail vast costs, they used the precise maps of the world created by the Google Street View car fleet.

Some of the achievements demonstrated an eerie human-like quality. For example, the car's vision system had the ability to recognize construction zones, slow down accordingly, and make its way through the cones safely. It also could adjust for vehicles partially blocking a lane, moving over as necessary. The system had not only been able to recognize bicyclists, but it could identify their hand signals and slow down to allow for them to change lanes in front. That suggested that Google was making progress on an even harder problem: What would a driverless car do when it was confronted by a cop making hand signal motions at an accident or a construction zone?

MIT roboticist John Leonard had taken particular joy in driving around Cambridge and shooting videos of the most confounding situations for autonomous vehicles. In one of his videos his car rolls up to a stop sign at a T-intersection and is waiting to make a left turn. The car is delayed by a long line of traffic passing from right to left, without a stop sign. The situation is complicated by light traffic coming from the opposite direction. The challenge is to persuade the drivers in the slow lane to give way, while not colliding with one of the cars zipping by at higher speed in the other direction.[7]

The video that was perhaps the toughest challenge for the Google vision system was taken at a busy crosswalk somewhere downtown. There is a crowd of people at a pedestrian crossing with a stoplight. The car is approaching when suddenly, completely ignoring the green light for the cars, a police officer's hand abruptly shoots out on the left of the frame to stop traffic for the pedestrians. Ultimately it may not be an impossible problem to solve with computer vision. If today's systems can already recognize cyclists and hand signals, uniforms cannot be far behind. But it will not be solved easily or quickly.

Intent on transforming education with massive open online courses, or MOOCs, and not wishing to compete for leadership of X Lab with Google cofounder Sergey Brin, Thrun largely departed the research program in 2012. As is often the case in Silicon Valley, Thrun had not been able to see his project through. After creating and overseeing the secret X Laboratory at Google for several years, he decided it was time for him to move on when Brin joined the effort. Brin proposed that they be codirectors, but Thrun realized that with Google's cofounder in the mix he would no longer be in control and so it was time for a new challenge.

In the fall of 2011 Thrun and Peter Norvig had taught one

of several free Stanford online course offerings, an Introduction to Artificial Intelligence. It made a big splash. More than 160,000 students signed up for the course, which was almost ten times the size of Stanford's nonvirtual student body. Although only a fraction of those enrolled in the course would ultimately complete it, it became a global "Internet moment": Thrun and Norvig's class raised the specter of a new low-cost form of education that would not only level the playing field by putting the world's best teachers within reach of anyone in the world, but also threaten the business models of high-priced elite universities. Why pay Stanford tuition if you could take the course anyway as a City College student?

Thrun was still nominally participating one day a week at Google, but the project leadership role was taken by Chris Urmson, a soft-spoken roboticist who had been Red Whittaker's chief lieutenant in the DARPA vehicle challenges. He had been one of the first people that Thrun hired after he came to Google to start the then secret car program. In the summer of 2014 he said he wanted to create a reliable driverless car before his son reached driving age, which was about six years in the future.

After Thrun departed, Urmson took the program a long way toward its original goal of autonomous driving on the open road. Google had divided the world into highway driving and driving in urban conditions. At a press conference called to summarize their achievements, Google acknowledged that their greatest challenge was figuring out how to program the car to drive in urban areas. Urmson, however, argued in a posting he made on the company's website that the chaos of the city streets with cars, bicyclists, and pedestrians moving in apparently random fashion, was actually reasonably predictable. The Google training experiment had encountered thousands of these situations and the company had developed software models that would expect both the expected (a car stopped at a red light) and the unexpected (a car running a red light). He and his team implied that the highway driving challenge was largely solved, with one caveat—the challenge of keeping the human driver engaged. That problem presented itself when the Google team farmed out some of their fleet of robotic vehicles to Google employees to test during their daily commute. "We saw some things that made us nervous," Urmson told a reporter. The original Google driving program had involved two professional drivers who worked from an aircraft-style checklist. The person in the driver's seat was vigilant and ready to take command if anything anomalous should happen. The real world was different. Some of the Google employees, on their way home after a long day at the office, had the disturbing habit of becoming distracted, up to and including falling asleep!

This was called the "handoff" problem. The challenge was to find a way to quickly bring a distracted human driver who might be reading email, watching a movie, or even sleeping back to the level of "situational awareness" necessary in an emergency. Naturally, people nodded off a lot more often in driverless cars that they had come to trust. It was something that the automotive industry would address in 2014 in the traffic jam assist systems that would drive cars in stop-and-go highway traffic. The drivers had to keep at least one hand on the wheel, except for ten-second intervals. If the driver didn't demonstrate "being there," the system gave an audible warning and took itself out of its self-driving mode. But automobile emergencies take place during a fraction of a second. Google decided that while in some distant future that might be a solvable problem, it wasn't possible to solve now with existing technology.

A number of other automakers are already attempting to deal with the problem of driver distraction. Lexus and Mercedes have commercialized technology that watches the driver's eyes and head position to determine if they are drowsy or distracted. Audi in 2014 began developing a system that would use two cameras to detect when a driver was inattentive and then bring the car to an abrupt halt, if needed.

For now, however, Google seems to have changed its strategy and is trying to solve another, perhaps simpler problem. In May of 2014, just weeks after they had given reporters an optimistic briefing on the progress of their driverless car, they shifted gears and set out to explore a new limited but more radical solution to autonomous transportation in urban environments. Unable to solve the distracted human problem, Google's engineers decided to take humans out of the loop entirely. The company de-emphasized its fleet of Prius and Lexus autonomous vehicles and set out to create a new fleet of a hundred experimental electric vehicles that entirely dispense with the standard controls in a modern automobile. Although it had successfully kept it a secret, Google had actually begun its original driverless vehicle program by experimenting with autonomous golf carts on the Google campus very early in the self-driving program. Now it was planning to return to its roots and once again autonomously ferry people around the Google campus, this time with its new specially designed vehicle fleet. Riding in the new Google car of the future will be more like riding in an elevator. The two-seat vehicle looks a bit like an ultracompact Fiat 500 or Mercedes-Benz Smart car, but with the steering wheel, gas pedal, brake, and gearshift all removed. The idea is that in crowded downtowns or on campuses passengers could enter the desired destination on their smartphones to summon a car on demand. Once they are inside, the car provides the passengers only a Trip Start button and a red E-Stop panic button. One of the conceptual shifts the engineers made was limiting the speed of the vehicle to just twenty-five miles per hour, allowing the Google cars to be regulated like golf carts rather than conventional automobiles. That meant that they could forgo air bags and other design restrictions that add cost, weight, and complexity. These limitations, however, mean that the new cars are suited only for low-speed urban driving.

Although 25 miles per hour is below highway standards, the average traffic speeds in San Francisco and New York are

18 and 17 miles per hour, respectively, and so it is possible that slow but efficient fleets of automated cars might one day replace today's taxis. A study by the Earth Institute found that Manhattan's 13,000 taxis make 470,000 trips a day. Their average speed is 10 to 11 miles per hour, carrying an average of 1.4 passengers an average distance of 2 miles, with a 5-minute average wait time to get a taxi. In comparison, the report said, it would be possible for a futuristic robot fleet of 9,000 automated vehicles hailed by smartphone to match that capacity with a wait time of less than 1 minute. Assuming a 15 percent profit, the current cost of today's taxi service would be about $4 per trip mile, while in contrast, it was estimated, a future driverless vehicle fleet would cost about 50 cents per mile. The report showed similar savings in two other case studies—in Ann Arbor, Michigan, and Babcock Ranch, a planned community in Florida.[8]

Google executives and engineers have made the argument that has long been advocated by urban planners: A vast amount of space is wasted on an automotive fleet that is rarely used. Cars used to commute, for example, sit parked for much of the day, taking up urban space that could be better used for housing, offices, or parks. In urban areas, automated taxis would operate continuously, only returning to a fast-charging facility for robots to swap out their battery packs. In this light, it is easy to reimagine cities not built around private cars, with more green space and broad avenues—which would be required to safely accommodate pedestrians and cyclists.

Thrun evoked both the safety issues and the potential to redesign cities when he spoke about the danger and irrationality of our current transportation system. In addition to squandering a great deal of resources, our transportation infrastructure is responsible for more than thirty thousand road deaths annually in the United States, and about ten times more than that in both India and China, which amounts to more than one million annual road deaths worldwide. It is a compelling argument, but it has been greeted with push-back both in terms of liability issues and more daunting ethical questions. An argument against autonomous vehicles is that the legal system is unequipped to sort out the culpability underpinning an accident that results from a design or implementation error. This issue speaks to the already incredibly complicated relationship between automotive design flaws and legal consequences. Toyota's struggle with claims of unintended acceleration, for example, cost the company more than $1.2 billion in damages. General Motors also grappled with a design flaw involving sudden stops because of a faulty ignition switch that resulted in the recall of more vehicles than they manufactured in 2014, and may ultimately cost several billion dollars. Yet there is potentially a simple remedy for this challenge. Congress could create a liability exemption for self-driving vehicles, as it has done for childhood vaccines. Insurance companies could impose a no-fault regime when only autonomous vehicles are involved in accidents.

Another aspect of the liability issue is what has been described as a version of the "trolley problem," which is generally stated thus: A runaway trolley is hurtling down the tracks toward five people who will be killed if it proceeds on its present course. You can save these five people by diverting the trolley onto a different set of tracks that has only one person on it, but that person will be killed. Is it morally permissible to turn the trolley and thus prevent five deaths at the cost of one? First posed as a thought problem in a paper about the ethics of abortion by British philosopher Philippa Foot in 1967, it has led to endless philosophical discussions on the implications of choosing the lesser evil.[9] More recently it has been similarly framed for robot vehicles deciding between avoiding five schoolchildren who have run out onto the road when the only option is swerving onto the sidewalk to avoid them, thus killing a single adult bystander.

Software could generally be designed to choose the lesser

evil; however, the framing of the question seems wrong on other levels. Because 90 percent of road accidents result from driver error, it is likely that a transition to autonomous vehicles will result in a dramatic drop in the overall number of injuries and deaths. So, clearly the greater good would be served even though there will still be a small number of accidents purely due to technological failures. In some respects, the automobile industry has already agreed with this logic. Air bags, for example, save more lives than are lost due to faulty air bag deployments.

Secondly, the narrow focus of the question ignores how autonomous vehicles will probably operate in the future, when it is highly likely that road workers, cops, emergency vehicles, cars, pedestrians, and cyclists will electronically signal their presence to each other, a feature that even without complete automation should dramatically increase safety. A technology known as V2X that continuously transmits the location of nearby vehicles to each other is now being tested globally. In the future, even schoolchildren will be carrying sensors to alert cars to their presence and reduce the chance of an accident.

It's puzzling, then, that the philosophers generally don't explore the trolley problem from the point of view of the greater good, but rather as an artifact of individual choice. Certainly it would be an individual tragedy if the technology fails—and of course it will fail. Systems that improve the overall safety of transportation seem vital, even if they aren't perfect. The more interesting philosophical conundrum is over the economic, social, and even cultural consequences of taking humans out of the loop in driving. More than 34,000 people died in 2013 in the United States in automobile accidents, and 2.36 million were injured. Balance that against the 3.8 million people who earned a living by driving commercially in the United States in 2012.[10] Driverless cars and trucks would potentially displace many if not most of those jobs as they emerge during the next two decades.

Indeed, the question is more nuanced than one narrowly posed as a choice of saving lives or jobs. When Doug Engelbart gave what would later be billed as "The Mother of All Demos" in 1968—a demonstration of the technologies that would lead to personal computing and the Internet—he implicitly adopted the metaphor of driving. He sat at a keyboard and a display and showed how graphical interactive computing could be used to control computing and "drive" through what would become known as cyberspace. The human was very much in control in this model of intelligence augmentation. Driving was the original metaphor for interactive computing, but today Google's vision has changed the metaphor. The new analogy will be closer to traveling in an elevator or a train without human intervention. In Google's world you will press a button and be taken to your destination. This conception of transportation undermines several notions that are deeply ingrained in American culture. In the last century the car became synonymous with the American ideal of freedom and independence. That era is now ending. What will replace it?

It is significant that Google is instrumental in changing the metaphor. In one sense the company began as the quintessential intelligence augmentation, or IA, company. The PageRank algorithm Larry Page developed to improve Internet search results essentially mined human intelligence by using the crowd-sourced accumulation of human decisions about valuable information sources. Google initially began by collecting and organizing human knowledge and then making it available to humans as part of a glorified Memex, the original global information retrieval system first proposed by Vannevar Bush in the *Atlantic Monthly* in 1945.[11]

As the company has evolved, however, it has started to push heavily toward systems that replace rather than extend humans. Google's executives have obviously thought to some degree about the societal consequences of the systems they are creating. Their corporate motto remains "Don't be evil."

Of course, that is nebulous enough to be construed to mean almost anything. Yet it does suggest that as a company Google is concerned with more than simply maximizing shareholder value. For example, Peter Norvig, a veteran AI scientist who has been director of research at Google since 2001, points to partnerships between human and computer as the way out of the conundrum presented by the emergence of increasingly intelligent machines. A partnership between human chess experts and a chess-playing computer program can outplay even the best AI chess program, he notes. "As a society that's what we're going to have to do. Computers are going to be more flexible and they're going to do more, and the people who are going to thrive are probably the ones who work in a partnership with machines," he told a NASA conference in 2014.[12]

What will the partnerships between humans and intelligent cars of the future look like? What began as a military plan to automate battlefield logistics, lowering costs and keeping soldiers out of harm's way, is now on the verge of reframing modern transportation. The world is plunging ahead and automating transportation systems, but the consequences are only dimly understood today. There will be huge positive consequences in safety, efficiency, and environmental quality. But what about the millions of people now employed driving throughout the world? What will they do when they become the twenty-first-century equivalent of the blacksmith or buggy-whip maker?

3 | A TOUGH YEAR FOR THE HUMAN RACE

"W"ith these machines, we can make any consumer device in the world," enthused Binne Visser, a Philips factory engineer who helped create a robot assembly line that disgorges an unending stream of electric shavers. His point is they could be smartphones, computers, or virtually anything that is made today by hand or using machines.[1]

The Philips electric razor factory in Drachten, a three-hour train ride north from Amsterdam through pancake-flat Dutch farmland, offers a clear view of the endgame of factory robots: that "lights-out," completely automated factories are already a reality, but so far only in limited circumstances. The Drachten plant feels from the outside like a slightly faded relic of an earlier era when Philips, which started out making lightbulbs and vacuum tubes, grew to be one of the world's dominant consumer electronics brands. Having lost its edge to Asian upstarts in consumer products such as television sets, Philips remains one of the world's leading makers of electric shavers and a range of other consumer products. Like many European and U.S. companies, it has based much of its manufacturing in Asia where labor is less expensive. A turning point came in 2012 when Philips scrapped a plan to move a high-end shaver assembly operation to China. Because of the falling prices of sensors, robots, and cameras and the increasing transportation costs to ship finished goods to markets outside Asia, Philips built an almost entirely automated assembly line of robot arms at the Drachten factory. Defeated in many consumer electronics categories, Philips decided to invest to maintain its edge in an eclectic array of precomputer home appliances.

The brightly lit single-story automated shaver factory is a modular mega-machine composed of more than 128 linked stations—each one a shining transparent cage connected to its siblings by a conveyor, resembling the glass-enclosed popcorn makers found in movie theaters. The manufacturing line itself is a vast Rube Goldberg–esque orchestra. Each of the 128 arms has a unique "end effector," a specialized hand for performing the same operation over and over and over again at two-second

A Philips shaver assembly plant in Drachten, Netherlands, that operates without assembly workers. (*Photo courtesy of Philips*)

intervals. One assembly every two seconds translates into 30 shavers a minute, 1,800 an hour, 1,304,000 a month, and an astounding 15,768,000 a year.

The robots are remarkably dexterous, each specialized to repeat its single task endlessly. One robot arm simultaneously picks up two toothpick-thin two-inch pieces of wire, precisely bends the wires, and then delicately places their stripped ends into tiny holes in a circuit board. The wires themselves are picked from a parts feeder called a shake table. A human technician loads them into a bin that then spills them onto a brightly lit surface observed by a camera placed overhead. As if playing Pick Up Sticks, the robot arm grabs two wires simultaneously. Every so often, when the wires are jumbled, it shakes the table to separate them so it can see them better and then quickly grabs two more. Meanwhile, a handful of humans flutter around the edges of the shaver manufacturing line. A team of engineers dressed in blue lab coats keeps the system running by feeding it raw materials. A special "tiger team" is on-call around the clock so no robot arm is ever down for more than two hours. Unlike human factory workers, the line never sleeps.

The factory is composed of American robot arms programmed by a team of European automation experts. Is it a harbinger of an era of manufacturing in which human factory line workers will vanish? Despite the fact that in China millions of workers labor to hand-assemble similar consumer gadgets, the Drachten plant is assembling devices more mechanically complex than a smartphone—entirely without human labor. In the automated factory mistakes are rare—the system is meant to be tolerant of small errors. At one station, toward the end of the line, small plastic pieces of the shaver case are snapped in place just beneath the rotary cutting head. One of the pieces, resembling a guitar pick, pops off onto the floor, like a Tiddlywink. The line doesn't stutter. A down-the-line sensor recognizes that the part is missing and the shaver is shunted aside

into a special rework area. The only humans directly working on the shaver factory line are eight women performing the
last step in the process: quality inspection, not yet automated
because the human ear is still the best instrument for deter-
mining that each shaver is functioning correctly.

Lights-out factories, defined as robotic manufacturing lines
without humans, create a "good news, bad news" scenario. To
minimize the total cost of goods, it makes sense to place factories either near sources of raw materials, labor, and energy
or near the customers for the finished goods. If robots can
build virtually any product more cheaply than human workers, then it is more economical for factories to be close to the
markets they serve, rather than near sources of low-cost labor.
Indeed, factories are already returning to the United States. A
solar panel manufacturing factory run by Flextronics has now
located in Milpitas, south of San Francisco, where a large banner proudly proclaims, BRINGING JOBS & MANUFACTURING BACK
TO CALIFORNIA! Walking the Fremont factory line, however, it
quickly becomes clear that the facility is a testament to highly
automated manufacturing rather than creating jobs; there
are fewer than ten workers actually handling products on the
assembly line producing almost as many panels as hundreds
of employees would in the company's conventional factory in
Asia. "At what point does the chainsaw replace Paul Bunyan?"
a Flextronics executive asks. "There's always a price point, and
we're very close to that point."[2]

At the dawn of the Information Age, the pace and consequences of automation were very much on Norbert Wiener's mind. During the summer of 1949, Wiener wrote a
single-spaced three-page letter to Walter Reuther, the head of
the United Auto Workers, to tell Reuther that he had turned
down a consulting opportunity with a General Electric corporation to offer technical advice on designing automated

machinery. GE had approached the MIT scientist twice in 1949 asking him both to lecture and to consult on the design of servomechanisms for industrial control applications. Servos used feedback to precisely control a component's position, which was essential for the automated machinery poised to enter the factory after World War II. Wiener had refused both offers for what he called ethical reasons, even though he realized that others with similar knowledge but no sense of obligation to factory workers would likely accept.

Wiener, deeply attuned to the potential dire "social consequences," had already unsuccessfully attempted to contact other unions, and his frustration came through clearly in the Reuther letter. By late 1942 it was clear to Wiener that a computer could be programmed to run a factory, and he worried about the ensuing consequences of an "assembly line without human agents."[3] Software had not yet become a force that, in the words of browser pioneer Marc Andreessen, would "eat the world," but Wiener portrayed the trajectory clearly to Reuther. "The detailed development of the machine for particular industrial purpose is a very skilled task, but not a mechanical task," he wrote. "It is done by what is called 'taping' the machine in the proper way, much as present computing machines are taped."[4] Today we call it "programming," and software animates the economy and virtually every aspect of modern society.

Writing to Reuther, Wiener foresaw an apocalypse: "This apparatus is extremely flexible, and susceptible to mass production, and will undoubtedly lead to the factory without employees; as for example, the automatic automobile assembly line," he wrote. "In the hands of the present industrial set-up, the unemployment produced by such plants can only be disastrous." Reuther responded by telegram: DEEPLY INTERESTED IN YOUR LETTER. WOULD LIKE TO DISCUSS IT WITH YOU AT EARLIEST OPPORTUNITY.

Reuther's response was sent in August 1949 but it was not until March 1951 that the two men met in a Boston hotel.[5] They sat together in the hotel restaurant and agreed to form a joint "Labor-Science-Education Association"[6] to attempt to deflect the worst consequences of the impending automation era for the nation's industrial workers. By the time Wiener met with Reuther he had already published *The Human Use of Human Beings*, a book that argued both for the potential benefits of automation and warned about the possibility of human subjugation by machines. He would become a sought-after national speaker during the first half of the 1950s, spreading his message of concern both about the possibility of runaway automation and the concept of robot weapons. After the meeting Wiener enthused that he had "found in Mr. Reuther and the men about him exactly that more universal union statesmanship which I had missed in my first sporadic attempts to make union contacts."[7]

Wiener was not the only one to attempt to draw Reuther's attention to the threat of automation. Several years after meeting with Wiener, Alfred Granakis, president of UAW 1250, also wrote to Reuther, warning him about the loss of jobs after he was confronted with new workplace automation technologies at a Ford Motor engine plant and foundry in Cleveland, Ohio. He described the plant as "today's nearest approach to a fully automated factory in the automobile industry," adding: "What is the economic solution to all this, Walter? I am greatly afraid of embracing an economic 'Frankenstein' that I helped create in its infancy. It is my opinion that troubled days lie ahead for Labor."[8]

Wiener had broken with the scientific and technical establishment some years earlier. He expressed strong beliefs about ethics in science in a letter to the *Atlantic Monthly* titled "A Scientist Rebels," published in December 1946, a year after he had suffered a crisis of conscience resulting from the bombing of Hiroshima and Nagasaki. The essay contained this response to a Boeing research scientist's request for a technical analysis of a guided missile system during the Second World War: "The practical use of guided missiles can only be used to kill foreign civilians indiscriminately, and it furnishes no protection whatever to civilians in this country."[9] The same letter raises the moral question of the dropping of the atomic bomb: "The interchange of ideas which is one of the great traditions of science must of course receive certain limitations when the scientist becomes an arbiter of life and death."[10]

In January of 1947 he withdrew from participating in a symposium on calculating machinery at Harvard University, in protest that the systems were to be used for "war purposes." In the 1940s both computers and robots were entirely the stuff of science fiction, so it's striking how clearly fleshed out Wiener's understandings were of the technology impact that is only today playing out. In 1949, the *New York Times* invited Wiener to summarize his views about "what the ultimate machine age is likely to be," in the words of its longtime Sunday editor, Lester Markel. Wiener accepted the invitation and wrote a draft of the article; the legendarily autocratic Markel was dissatisfied and asked him to rewrite it. He did. But through a distinctly pre-Internet series of fumbles and missed opportunities, neither version ever appeared at the time.

In August of 1949, according to Wiener's papers at MIT, the *Times* asked him to resend the first draft of the article to be combined with the second draft. (It is unclear why the editors had misplaced the first draft.) "Could you send the first draft to me, and we'll see whether we can combine the two into one story?" wrote an editor in the paper's Sunday department, then separate from the daily paper. "I may be mistaken, but I think you lost some of your best material." But by then Wiener was traveling in Mexico, and he responded: "I had assumed that the first version of my article was finished business. To get hold of the paper in my office at the Massachusetts Institute of Technology would involve considerable cross-correspondence and annoyance to several people. I therefore do not consider it a practical thing to do. Under the circumstances I think that it is best for me to abandon this undertaking."

The following week the *Times* editor returned the second draft to Wiener, and it eventually ended up with his papers in MIT Libraries' Archives and Special Collections, languishing there until December 2012, when it was discovered by Anders Fernstedt, an independent scholar researching the work of Karl Popper and Friedrich Hayek, three Viennese philosophers, and Ernst Gombrich, an art historian, active in London for most of the twentieth century.[11] In the unpublished essay Wiener's reservations were clear: "The tendency of these new machines is to replace human judgment on all levels but a fairly high one, rather than to replace human energy and power by machine energy and power. It is already clear that this new replacement will have a profound influence upon our lives," he wrote.

Wiener went on to mention the emergence of factories that were "substantially without employees" and the rise of the importance of "taping." He also presented more than a glimmer of the theoretical possibility and practical impact of machine learning: "The limitations of such a machine are simply those of an understanding of the objects to be attained, and of the potentialities of each stage of the processes by which they are to be attained, and of our power to make logically determinate combinations of those processes to achieve our ends. Roughly speaking, if we can do anything in a clear and intelligible way, we can do it by machine."[12]

At the dawn of the computer age, Wiener could see and clearly articulate that automation had the potential of reducing the value of a "routine" factory employee to where "he is not worth hiring at any price," and that as a result "we are in for an industrial revolution of unmitigated cruelty."

Not only did he have early dark forebodings of the computer revolution, but he foresaw something else that was even more chilling: "If we move in the direction of making machines which learn and whose behavior is modified by experience, we must face the fact that every degree of independence we give the machine is a degree of possible defiance of our wishes. The genie in the bottle will not willingly go back in the bottle, nor have we any reason to expect them to be well disposed to us."[13]

In the early 1950s Reuther and Wiener agreed on the idea of a "Labor-Science-Education Association," but the partnership did not have an immediate impact, in part because of Wiener's health issues and in part because Reuther represented a faction of the U.S. labor movement that viewed automation as unavoidable progress—the labor leader was intent on forging an economic bargain with management around the forces of technology: "In the final analysis, modern work processes had to be endured, offset by the reward of increased leisure and creative relaxation. In his embrace of automation and new technology, he often seemed to be wholly taken by the notion of efficiency as a desirable and essentially neutral condition."[14]

Wiener's warning would eventually light a spark—but not during the 1950s, a Republican decade when the labor movement did not have many friends in government. Only after Kennedy's election in 1960 and his succession by Lyndon Johnson would the early partnership between Wiener and Reuther lead to one of the few serious efforts on the part of the U.S. government to grapple with automation, when in August of 1964 Johnson established a blue-ribbon panel to explore the impact of technology on the economy.

Pressure came in part from the Left in the form of an open letter to the president from a group that called itself the Ad Hoc Committee on the Triple Revolution, including Michael Harrington, who would later help found the Democratic Socialists of America; Students for a Democratic Society cofounder Tom Hayden; chemist Linus Pauling; Swedish economist Gunnar Myrdal; pacifist A. J. Muste; economic historian Robert Heilbroner; social critic Irving Howe; civil rights activist Bayard Rustin; and Socialist Party presidential candidate Norman Thomas, among many others.

The first revolution they noted was the emergence of the "Cybernation": "A new era of production has begun. Its prin-

ciples of organization are as different from those of the industrial era as those of the industrial era were different from the
agricultural. The cybernation revolution has been brought
about by the combination of the computer and the automated
self-regulating machine. This results in a system of almost
unlimited productive capacity which requires progressively
less human labor."[15] The resulting National Commission on
Technology, Automation, and Economic Progress would include
a remarkable group ranging from Reuther, Thomas J. Watson
Jr. of IBM, and Edwin Land of Polaroid, to Robert Solow, the
MIT economist, and Daniel Bell, the Columbia sociologist.

When the 115-page report appeared at the end of 1966 it
was accompanied by 1,787 pages of appendices including special reports by outside experts. The 232-page analysis of the
impact of computing by Paul Armer of the RAND Corporation
did a remarkable job of predicting the impact of information
technology. Indeed, the headings in the report have proven
true over the years: "Computers Are Becoming Faster, Smaller,
and Less Expensive"; "Computing Power Will Become Available Much the Same as Electricity and Telephone Service Are
Today"; "Information Itself Will Become Inexpensive and Readily Available"; "Computers Will Become Easier to Use"; "Computers Will Be Used to Process Pictorial Images and Graphic
Information"; and "Computers Will Be Used to Process Language," among others. Yet the consensus that emerged from
the report would be the traditional Keynesian view that "technology eliminated jobs, not work." The report concluded that
technological displacement would be a temporary but necessary stepping-stone for economic growth.

The debate over the future of technological unemployment
dissipated as the economy heated up, in part as a consequence
of the Vietnam War, and the postwar civil strife in the late
1960s further sidelined the question. A decade and a half after
he had issued his first warnings about the consequences of
automated machines, Wiener turned his thoughts to religion

and technology while remaining a committed humanist. In his final book, *God & Golem, Inc.,* he explored the future human relationship with machines through the prism of religion. Invoking the parable of the golem, he pointed out that despite best intentions, humans are incapable of understanding the ultimate consequences of their inventions.[16]

In his 1980 dual biography of John von Neumann and Wiener, Steven Heims notes that in the late 1960s he had asked a range of mathematicians and scientists about Wiener's philosophy of technology. The general reaction of the scientists was as follows: "Wiener was a great mathematician, but he was also eccentric. When he began talking about society and the responsibility of scientists, a topic outside of his area of expertise, well, I just couldn't take him seriously."[17]

Heims concludes that Wiener's social philosophy hit a nerve with the scientific community. If scientists acknowledged the significance of Wiener's ideas, they would have to reexamine their deeply held preconceived notions about personal responsibility, something they were not eager to do. "Man makes man in his own image," Wiener notes in *God and Golem, Inc.* "This seems to be the echo or the prototype of the act of creation, by which God is supposed to have made man in His image. Can something similar occur in the less complicated (and perhaps more understandable) case of the nonliving systems that we call machines?"[18]

Shortly before his death in 1964, Wiener was asked by *U.S. News & World Report*: "Dr. Wiener, is there any danger that machines—that is, computers—will someday get the upper hand over men?" His answer was: "There is, definitely, that danger if we don't take a realistic attitude. The danger is essentially intellectual laziness. Some people have been so bamboozled by the word 'machine' that they don't realize what can be done and what cannot be done with machines—and what can be left, and what cannot be left to the human beings."[19]

Only now, six and a half decades after Wiener wrote *Cyber-*

netics in 1948, is the machine autonomy question becoming more than hypothetical. The Pentagon has begun to struggle with the consequences of a new generation of "brilliant" weapons,[20] while philosophers grapple with the "trolley problem" in trying to assign moral responsibility for self-driving cars. Over the next decade the consequences of creating autonomous machines will appear more frequently as manufacturing, logistics, transportation, education, health care, and communications are increasingly directed and controlled by learning algorithms rather than humans.

Despite Wiener's early efforts to play a technological Paul Revere, after the automation debates of the 1950s and 1960s tailed off, fears of unemployment caused by technology would vanish from the public consciousness until sometime around 2011. Mainstream economists generally agreed on what they described as the "Luddite fallacy." As early as 1930, John Maynard Keynes had articulated the general view on the broad impact of new technology: "We are being afflicted with a new disease of which some readers may not yet have heard the name, but of which they will hear a great deal in the years to come—namely, *technological unemployment*. This means unemployment due to our discovery of means of economizing the use of labor outrunning the pace at which we can find new uses for labor. But this is only a temporary phase of maladjustment."[21]

Keynes was early to point out that technology was a powerful generator of new categories of employment. Yet what he referred to as a "temporary phase" is certainly relative. After all, he also famously noted that in "the long run" we are all dead.

In 1995, economist Jeremy Rifkin wrote *The End of Work: The Decline of the Global Labor Force and the Dawn of the Post-Market Era*. The decline of the agricultural economy and the rapid growth of new industrial employment had been a stunning substantiation of Keynes's substitution argument, but Rifkin argued that the impact of new information technologies would be qualitatively different from that of previous waves of industrial automation. He began by noting that in 1995 unemployment globally had risen to its highest level since the depression of the 1930s and that globally eight hundred million people were unemployed or underemployed. "The restructuring of production practices and the permanent replacement of machines for human laborers has begun to take a tragic toll on the lives of millions of workers," he wrote.[22]

The challenge to his thesis was that employment in the United States actually grew from 115 million to 137 million during the decade following the publication of his book. That meant that the size of the workforce would grow by over 19 percent while the nation's population grew by only 11 percent. Moreover, key economic indicators such as the labor force participation rate, employment to working population ratio, and the unemployment rate showed no evidence of technological unemployment. The situation, then, was more nuanced than the impending black-and-white labor calamity Rifkin had forecast. For example, from the 1970s, the outsourcing of jobs internationally, as multinational corporations fled to low-cost manufacturing regions and used telecommunications networks to relocate white-collar jobs, had a far more significant impact on domestic employment than the deployment of automation technologies. And so Rifkin's work, substantially discredited, also went largely unnoticed.

In the wake of the 2008 recession, there were indications of a new and broader technology transformation. White-collar employment had been the engine of growth for the U.S. economy since the end of World War II, but now cracks began to appear. What were once solid white-collar jobs began disappearing. Routinized white-collar work was now clearly at risk as the economy began to recover in 2009 in the form of what was described as a "jobless recovery." Indications were that knowledge workers' jobs higher up in the economic pyramid were for the first time vulnerable. Economists such as MIT's

David Autor began to pick apart the specifics of the changing labor force and put forward the idea that the U.S. economy was being "hollowed out." It might continue to grow at the bottom and the top, but middle-class jobs, essential to a modern democracy, were evaporating, he argued.

There was mounting evidence that the impact of technology was not just a hollowing out but a "dumbing down" of the workforce. In some cases specific high-prestige professions began to show the impact of automation based on the falling costs of information and communications technologies, such as new global computer networks. Moreover, for the first time artificial intelligence software was beginning to have a meaningful impact on certain highly skilled jobs, like $400-per-hour lawyers and $175-per-hour paralegals. As the field of AI once again gathered momentum beginning in 2000, new applications of artificial intelligence techniques based on natural language understanding, such as "e-discovery," or the automated processing of the relevance of legal documents required to disclose in litigation, emerged. The software would soon go beyond just finding specific keywords in email. E-discovery software evolved quickly, so that it became possible to scan millions of documents electronically and recognize underlying concepts and even find so-called smoking guns—that is, evidence of illegal or improper behavior.

In part, the software had become essential as litigation against corporations routinely involved the review of millions of documents for relevance. Comparative studies showed that the machines could do as well or better than humans in analyzing and classifying documents. "From a legal staffing viewpoint, it means that a lot of people who used to be allocated to conduct document review are no longer able to be billed out," said Bill Herr, who as a lawyer at a major chemical company used to muster auditoriums of lawyers to read documents and correspondence for weeks on end. "People get bored, people get headaches. Computers don't."[23]

Observing the impact of technologies such as e-discovery software, which is now dramatically eliminating the jobs of lawyers, led Martin Ford, an independent Silicon Valley engineer who owned a small software firm, to self-publish *The Lights in the Tunnel: Automation, Accelerating Technology and the Economy of the Future* at the end of 2009. Ford had come to believe that the impact of information technology on the job market was moving much more quickly than was generally understood. With a professional understanding of software technologies, he was also deeply pessimistic. For a while he stood alone, much in the tradition of Rifkin's 1995 *The End of Work,* but as the recession dragged on and mainstream economists continued to have trouble explaining the absence of job growth, he was soon joined by an insurgency of technologists and economists warning that technological disruption was happening full force.

In 2011, two MIT Sloan School economists, Erik Brynjolfsson and Andrew McAfee, self-published an extended essay titled "Race Against the Machine: How the Digital Revolution Is Accelerating Innovation, Driving Productivity, and Irreversibly Transforming Employment and the Economy." Their basic theme was as follows: "Digital technologies change rapidly, but organizations and skills aren't keeping pace. As a result, millions of people are being left behind. Their incomes and jobs are being destroyed, leaving them worse off . . . than before the digital revolution."[24] The "Race Against the Machine" essay was passed around samizdat-style over the Internet and was instrumental in reigniting the debate over automation. The basic theme of the discussion was around the notion that this time—because of the acceleration of computing technologies in the workplace—there would be no Keynesian solution in which the economy created new job categories.

Like Martin Ford, Brynjolfsson and McAfee chronicled a growing array of technological applications that were redefining the workplace, or seemed poised on the brink of doing so. Of the wave of new critiques, David Autor's thesis was perhaps the most compelling. However, even he began to hedge in 2014, based on a report that indicated a growing "deskilling" of the U.S. workforce and a declining demand for jobs that required cognitive skills. He worried that the effect was creating a downward ramp. The consequence, argued Paul Beaudry, David A. Green, and Ben Sand in a National Bureau of Economic Research (NBER) working paper, was that higher-skilled workers tended to push lower-skilled workers out of the workforce.[25] Although they have no clear evidence directly related to the deployment of particular types of technologies, the analysis of the consequences for the top of the workforce is chilling. They reported: "Many researchers have documented a strong, ongoing increase in the demand for skills in the decades leading up to 2000. In this paper, we document a decline in that demand in the years since 2000, even as the supply of high education workers continues to grow. We go on to show that, in response to this demand reversal, high-skilled workers have moved down the occupational ladder and have begun to perform jobs traditionally performed by lower-skilled workers."[26] Yet despite fears of a "job apocalypse" based on machines that can see, hear, speak, and touch, once again the workforce has not behaved as if there will be a complete collapse precipitated by technological advance in the immediate future. Indeed, in the decade from 2003 to 2013, the size of the U.S. workforce increased by more than 5 percent, from 131.4 million to 138.3 million—although, to be sure, this was a period during which the population grew by more than 9 percent.

If not complete collapse, the slowing growth rate suggested a more turbulent and complex reality. One possibility is that rather than a pure deskilling, the changes observed may represent a broader "skill mismatch," an interpretation that is more consistent with Keynesean expectations. For example, a recent McKinsey report on the future of work showed that between 2001 and 2009, jobs related to transactions and production both declined, but more than 4.8 million white-collar jobs were created relating to interactions and problem-solving.[27] What is clear is that both blue-collar and white-collar jobs involving routinized tasks are at risk. The *Financial Times* reported in 2013 that between 2007 and 2012 the U.S. workforce gained 387,000 managers while losing almost two million clerical jobs.[28] This is an artifact of what is popularly described as the Web 2.0 era of the Internet. The second generation of commercial Internet applications brought the emergence of a series of software protocols and product suites that simplified the integration of business functions. Companies such as IBM, HP, SAP, PeopleSoft, and Oracle, helped corporations to relatively quickly automate repetitive business functions. The consequence has been a dramatic loss of clerical jobs.

However, even within the world of clerical labor there are subtleties that suggest that predictions of automation and job destruction across the board are unlikely to prove valid. The case of bank tellers and the advent of automated teller machines is a particularly good example of the complex relationship between automation technologies, computer networks, and workforce dynamics. In 2011, while discussing the economy, Barack Obama used this same example: "There are some structural issues with our economy where a lot of businesses have learned to become much more efficient with a lot fewer workers. You see it when you go to a bank and you use an ATM; you don't go to a bank teller. Or you go to the airport, and you're using a kiosk instead of checking in at the gate."[29]

This touched off a political kerfuffle about the impact of automation. The reality is that despite the rise of ATMs, bank tellers have not gone away. In 2004 Charles Fishman reported in *Fast Company* that in 1985, relatively early in the deployment of ATMs, there were about 60,000 ATMs and 485,000 bank tellers; in 2002 that number had increased to 352,000 ATMs and 527,000 bank tellers. In 2011 the *Economist* cited 600,500 bank tellers in 2008, while the Bureau of Labor Statistics was projecting that number would grow to 638,000 by 2018. Furthermore the *Economist* pointed out that there were an additional 152,900 "computer, automated teller, and office machine repairers" in 2008.[30] Focusing on ATMs in isolation doesn't begin to touch the complexity of the way in which automated systems are weaving their way into the economy.

Bureau of Labor Statistics data reveal that the real transformation has been in the "back office," which in 1972 made up 70 percent of the banking workforce: "First, the automation of a major customer service task reduced the number of employees per location to 75% of what it was. Second, the [ATM] machines did not replace the highly visible customer-facing bank tellers, but instead eliminated thousands of less-visible clerical jobs."[31] The impact of back-office automation in banking is difficult to estimate precisely, because the BLS changed the way it recorded clerk jobs in banking in 1982. However, it is indisputable that banking clerks' jobs have continued to vanish.

Looking forward, the consequences of new computing technology on bank tellers might anticipate the impact of driverless delivery vehicles. Even if the technology can be perfected—and that is still to be determined, because delivery involves complex and diverse contact with human business and residential customers—the "last mile" delivery personnel will be hard to replace.

Despite the challenges of separating the impact of the recession from the implementation of new technologies, increasingly the connection between new automation technologies and rapid economic change has been used to imply that a collapse of the U.S. workforce—or at least a prolonged period of dislocation—might be in the offing. Brynjolfsson and McAfee argue for the possibility in a much expanded book-length version of "Race Against the Machine," entitled *The Second Machine Age: Work, Progress, and Prosperity in a Time of Brilliant Technologies.* Similar sentiments are offered by Jaron Lanier, a well-known computer scientist now at Microsoft Research, in the book *Who*

Owns the Future? Both books draw a direct link between the rise of Instagram, the Internet photo-sharing service acquired by Facebook for $1 billion in 2012, and the decline of Kodak, the iconic photographic firm that declared bankruptcy that year. "A team of just fifteen people at Instagram created a simple app that over 130 million customers use to share some sixteen billion photos (and counting)," wrote Brynjolfsson and McAfee. "But companies like Instagram and Facebook employ a tiny fraction of the people that were needed at Kodak. Nonetheless, Facebook has a market value several times greater than Kodak ever did and has created at least seven billionaires so far, each of whom has a net worth ten times greater than [Kodak founder] George Eastman did."[32]

Lanier makes the same point about Kodak's woes even more directly: "They even invented the first digital camera. But today Kodak is bankrupt, and the new face of digital photography has become Instagram. When Instagram was sold to Facebook for a billion dollars in 2012, it employed only thirteen people. Where did all those jobs disappear to? And what happened to the wealth that those middle-class jobs created?"[33]

The flaw in their arguments is that they mask the actual jobs equation and ignore the reality of Kodak's financial turmoil. First, even if Instagram did actually kill Kodak—it didn't—the jobs equation is much more complex than the cited 13 versus 145,000 disparity. Services like Instagram didn't spring up in isolation, but were made possible after the Internet had reached a level of maturity that had by then created millions of mostly high-quality new jobs. That point was made clearly by Tim O'Reilly, the book publisher and conference organizer: "Think about it for a minute. Was it really Instagram that replaced Kodak? Wasn't it actually Apple, Samsung, and the other smartphone makers who have replaced the camera? And aren't there network providers, data centers, and equipment suppliers who provide the replacement for the film that Kodak once sold? Apple has 72,000 employees (up from

10,000 in 2002). Samsung has 270,000 employees. Comcast has 126,000. And so on."[34] And even O'Reilly's point doesn't begin to capture the positive economic impact of the Internet. A 2011 McKinsey study reported that globally the Internet created 2.6 new jobs for every job lost, and that it had been responsible for 21 percent of GDP growth in the five previous years in developed countries.[35] The other challenge for the Kodak versus Instagram argument is that while Kodak suffered during the shift to digital technologies, its archrival FujiFilm somehow managed to prosper through the transition to digital.[36]

The reason for Kodak's decline was more complex than "they missed digital" or "they failed to buy (or invent) Instagram." The problems included scale, age, and abruptness. The company had a massive burden of retirees and an internal culture that lost talent and could not attract more. It proved to be a perfect storm. Kodak tried to get into pharmaceuticals in a big way but failed, and it failed in its effort to enter the medical imaging business.

The new anxiety about AI-based automation and the resulting job loss may eventually prove well founded, but it is just as likely that those who are alarmed have in fact just latched onto the right backward-facing snapshots. If the equation is framed in terms of artificial intelligence–oriented technologies versus those oriented toward augmenting humans, there is hope that humans still retain an unbounded ability to both entertain and employ themselves doing something marketable and useful.

If the humans are wrong, however, 2045 could be a tough year for the human race.

Or it could mark the arrival of a technological paradise.

Or both.

The year 2045 is when Ray Kurzweil predicts humans will transcend biology, and implicitly, one would presume, destiny.[37]

Kurzweil, the serial artificial intelligence entrepreneur and author who joined Google as a director of engineering in 2012 to develop some of his ideas for building an artificial "mind," represents a community of many of Silicon Valley's best and brightest technologists. They have been inspired by the ideas of computer scientist and science-fiction author Vernor Vinge about the inevitability of a "technological singularity" that would mark the point in time at which machine intelligence will surpass human intelligence. When he first wrote about the idea of the singularity in 1993, Vinge framed a relatively wide span of years—between 2005 and 2030—during which computers might become "awake" and superhuman.[38]

The singularity movement depends on the inevitability of mutually reinforcing exponential improvements in a variety of information-based technologies ranging from processing power to storage. In one sense it is the ultimate religious belief in the power of technology-driven exponential curves, an idea that has been explored by Robert Geraci in *Apocalyptic AI: Visions of Heaven in Robotics, Artificial Intelligence, and Virtual Reality.* There he finds fascinating sociological parallels between singularity thinking and a variety of messianic religious traditions.[39]

The singularity hypothesis also builds on the emergent AI research pioneered by Rodney Brooks, who first developed a robotics approach based on building complex systems out of collections of simpler parts. Both Kurzweil in *How to Create a Mind: The Secret of Human Thought Revealed* and Jeff Hawkins in his earlier *On Intelligence: How a New Understanding of the Brain Will Lead to the Creation of Truly Intelligent Machines* attempt to make the case that because the simple biological "algorithms" that are the basis for human intelligence have been discovered, it is largely a matter of "scaling up" to engineer intelligent machines. These ideas have been tremendously controversial and have been criticized by neuroscientists, but are worth mentioning here because they are an underlying argument in the new automation debate. What is most striking today is the extreme range of opinions about the future of the workforce emerging from different interpretations of the same data.

Moshe Vardi is a Rice University computer scientist who serves as editor-in-chief of the Communications of the Association for Computing Machinery. In 2012 he began to argue publicly that the rate of acceleration in AI was now so rapid that all human labor will become obsolete within just over three decades. In an October 2012 *Atlantic* essay, "The Consequences of Human Intelligence,"[40] Vardi took a position that is becoming increasingly representative of the AI research community: "The AI Revolution, however, is different, I believe, than the Industrial Revolution. In the 19th century machines competed with human brawn. Now machines are competing with human brain. Robots combine brain and brawn. We are facing the prospect of being completely out-competed by our own creations."[41]

Vardi believes that the areas where new job growth is robust—for example, in the Web search engine economy, where new categories of workers such as those who perform tasks like search engine optimization, or SEO—are inherently vulnerable in the very near term. "If I look at search engine optimization, yes, right now they are creating jobs in doing this," he said. "But what is it? It is learning how search engines actually work and then applying this to the design of Web pages. You could say that is a machine-learning problem. Maybe right now we need humans, but these guys [software automation designers] are making progress."[42]

The assumption of many like Vardi is that a market economy will not protect a human labor force from the effects of automation technologies. Like many of the "Singularitarians," he points to a portfolio of social engineering options for softening the impact. Brynjolfsson and McAfee in *The Second Machine Age* sketch out a broad set of policy options that have the flavor of a new New Deal, with examples like "teach the children well," "support our scientists," "upgrade infrastructure." Others like Harvard Business School professor Clayton Christensen have argued for focusing on technologies that create rather than destroy jobs (a very clear IA versus AI position).

At the same time, while many who believe in accelerating change agonize about its potential impact, others have a more optimistic perspective. In a series of reports issued beginning in 2013, the International Federation of Robotics (IFR), established in 1987 with headquarters in Frankfurt, Germany, self-servingly argued that manufacturing robots actually increased economic activity and therefore, instead of causing unemployment, both directly and indirectly increased the total number of human jobs. One February 2013 study claims the robotics industry would directly and indirectly create 1.9 million to 3.5 million jobs globally by 2020.[43] A revised report the following year argued that for every robot deployed, 3.6 jobs were created.

But what if the Singularitarians are wrong? In the spring of 2012 Robert J. Gordon, a self-described "grumpy" Northwestern University economist rained on the Silicon Valley "innovation creates jobs and progress" parade by noting that the claims for gains did not show up in conventional productivity figures. In a widely cited National Bureau of Economic Research white paper in 2012 he made a series of points contending that the productivity bubble in the twentieth century was a one-time event. He also noted that the automation technologies cited by those he would later describe as "techno-optimists" had not had the same kind of productivity impact as earlier nineteenth-century industrial innovations. "The computer and Internet revolution (IR3) began around 1960 and reached its climax in the dot-com era of the late 1990s, but its main impact on productivity has withered away in the past eight years," he wrote. "Many of the inventions that replaced tedious and repetitive clerical labour with computers happened a long time ago, in the 1970s and 1980s. Invention since 2000 has centered on entertainment and communication devices that are smaller, smarter, and more capable, but do not fundamentally change labour productivity or the standard of living in the way that electric light, motor cars, or indoor plumbing changed it."[44]

In one sense it was a devastating critique of the Silicon Valley faith in "trickle down" from exponential advances in integrated circuits, for if the techno-optimists were correct, the impact of new information technology should have resulted in a dramatic explosion of new productivity, particularly after the deployment of the Internet. Gordon pointed out that unlike the earlier industrial revolutions, there has not been a comparable productivity advance tied to the computing revolution. "They remind us Moore's Law predicts endless exponential growth of the performance capability of computer chips, without recognizing that the translation from Moore's Law to the performance-price behavior of ICT equipment peaked in 1998 and has declined ever since," he noted in a 2014 rejoinder to his initial paper.[45]

Gordon squared off with his critics, most notably with MIT economist Erik Brynjolfsson, at the TED Conference in the spring of 2013. In a debate moderated by TED host Chris Anderson, the two jousted over the future impact of robotics and whether the supposed exponentials would continue or were rather the peak of an "S curve" with a decline on the way.[46] The techno-optimists believe that a lag between invention and adoption of technology simply delays the impact of productivity gains and even though exponentials inevitably taper off, they spawn successor inventions—for example the vacuum tube was followed by the transistor, which in turn was followed by the integrated circuit.

Gordon, however, has remained a consistent thorn in the side of the Singularitarians. In a *Wall Street Journal* column, he asserted that there are actually relatively few productivity opportunities in driverless cars. Moreover, he argued, they will not have a dramatic impact on safety either—auto fatalities per miles traveled have already declined by a factor of ten since 1950, making future improvements less significant.[47] He also cast a skeptical eye on the notion that a new generation of mobile robots would make inroads into both the manufacturing and service sectors of the economy: "This lack of multitasking ability is dismissed by the robot enthusiasts—just wait, it is coming. Soon our robots will not only be able to win at *Jeopardy!* but also will be able to check in your bags at the skycap station at the airport, thus displacing the skycaps. But the physical tasks that humans can do are unlikely to be replaced in the next several decades by robots. Surely multiple-function robots will be developed, but it will be a long and gradual process before robots outside of the manufacturing and wholesaling sectors become a significant factor in replacing human jobs in the service or construction sectors."[48]

His skepticism unleashed a torrent of criticism, but he has refused to back down. His response to his critics is, in effect, "Be careful what you wish for!" Gordon has also pointed out that Norbert Wiener may have had the most prescient insight into the potential impact of the "Third Industrial Revolution" (IR3), of computing and the Internet beginning in about 1960, when he argued that automation for automation's sake would have unpredictable and quite possibly negative consequences.

The productivity debate has continued unabated. It has recently become fashionable for technologists and economists to argue that the traditional productivity benchmarks are no longer appropriate for measuring an increasingly digitized economy in which information is freely shared. How do you measure the economic value of a resource like Wikipedia, they ask? If the Singularitarians are right, however, the transformation in the form of an unparalleled economic crisis as human labor becomes surplus should be obvious soon. Indeed, the outcome might be quite gloomy; there will be fewer and fewer places for humans in the resulting economy.

That has certainly not happened yet in the industrialized world. However, one intriguing shift that suggests there are limits to automation was the recent decision by Toyota to systematically put working humans back into the manufacturing process. In quality and manufacturing on a mass scale, Toyota has been a global leader in automation technologies based on the corporate philosophy of *kaizen* (Japanese for "good change") or continuous improvement. After pushing its automation processes toward lights-out manufacturing, the company realized that automated factories do not improve themselves. Once Toyota had extraordinary craftsmen that were known as *Kami-sama,* or "gods" who had the ability to make anything, according to Toyota president Akio Toyoda.[49] The craftsmen also had the human ability to act creatively and thus improve the manufacturing process. Now, to add flexibility and creativity back into their factories, Toyota chose to restore a hundred "manual-intensive" workspaces.

The restoration of the Toyota gods is evocative of Stewart Brand's opening line to the 1968 *Whole Earth Catalog:* "We are as gods and might as well get good at it." Brand later acknowledged that he had borrowed the concept from British anthropologist Edmund Leach, who wrote, also in 1968: "Men have become like gods. Isn't it about time that we understood our divinity? Science offers us total mastery over our environment and over our destiny, yet instead of rejoicing we feel deeply afraid. Why should this be? How might these fears be resolved?"[50]

Underlying both the acrimonious productivity debate and Toyota's rebalancing of craft and automation is the deeper question about the nature of the relationship between humans and smart machines. The Toyota shift toward a more cooperative relationship between human and robot might alternatively suggest a new focus on technology for augmenting humans rather than displacing them. Singularitarians, however, argue that such human-machine partnerships are simply an interim stage during which human knowledge is transferred and at some point creativity will be transferred to or will even arise on its own in some future generation of brilliant machines. They point to small developments in the field of machine learning that suggest that computers will exhibit humanlike learning skills at some point in the not-too-distant future. In 2014, for example, Google paid $650 million to acquire DeepMind Technologies, a small start-up with no commercial products that had shown machine-learning algorithms with the ability to play video games, in some cases better than humans. When the acquisition was first reported it was rumored that because of the power and implications of the technology Google would set up an "ethics board" to evaluate any unspecified "advances."[51] It has remained unclear whether such oversight will be substantial or whether it was just a publicity stunt to hype the acquisition and justify its price.

It is undeniable that AI and machine-learning algorithms have already had world-transforming application in areas as diverse as science, manufacturing, and entertainment. Examples range from machine vision and pattern recognition essential in improving quality in semiconductor design and so-called rational drug discovery algorithms, which systematize the creation of new pharmaceuticals, to government surveillance and social media companies whose business model is invading privacy for profit. The optimists hope that potential abuses will be minimized if the applications remain human-focused rather than algorithm-centric. The reality is that, until now, Silicon Valley has not had a track record that is morally superior to any earlier industries. It will be truly remarkable if any Silicon Valley company actually rejects a profitable technology for ethical reasons.

Setting aside the philosophical discussion about self-aware machines, and in spite of Gordon's pessimism about productivity increases, it is clearly becoming increasingly possible and "rational" to design humans out of systems for both performance and cost reasons. Google, which can alternatively be seen as either an IA or AI company, seems to be engaged in an internal tug-of-war over this dichotomy. The original PageRank algorithm that the company is based on can perhaps be construed as the most powerful example in the history of human augmentation. The algorithm systematically mined human decisions about the value of information and pooled and ranked those decisions to prioritize Web search results. While some have chosen to criticize this as a systematic way to siphon intellectual value from vast numbers of unwitting humans, there is clearly an unstated social contract between user and company. Google mines the wealth of human knowledge and returns it to society, albeit with a monetization "catch." The Google search dialog box has become the world's most powerful information monopoly.

Since then, however, Google has yo-yoed back and forth in designing both IA and AI applications and services, whichever works best to solve the problem at hand. For example, for all of the controversy surrounding it, the Google Glass reality augmentation system clearly has the potential to be what the name promises—a human augmentation tool—while the Google car project represents the pros and cons of a pure AI system replacing human agency and intelligence with a machine. Indeed, Google as a company has become a de facto experiment about the societal consequences of AI-based technologies deployed on a massive scale. In a 2014 speech to the group of NASA scientists, Peter Norvig, Google's director of research, was clear that the only reasonable solution to AI advances would lie in designing systems in which humans partner with intelligent machines. His solution was a powerful declaration of intent about the need to converge the separate AI and IA communities.

Given the current rush to build automated factories, such a convergence seems unlikely on a broad societal basis. However, the dark fears that have surfaced recently about job-killing manufacturing robots are perhaps likely to soon be supplanted by a more balanced view of our relationship with machines beyond the workplace. Consider Terry Gou, the chief executive of Foxconn, one of the largest Chinese manufacturers and makers of the Apple iPhone. The company had already endured global controversy for labor conditions in its factories when, at the beginning of 2012, Gou declared that Foxconn was now planning a significant commitment to robots to replace his workers. "As human beings are also animals, to manage one million animals gives me a headache," he said during a business meeting.[52]

Although the statement drew global attention, his vision of a factory without workers is only one of the ways in which robotics will transform society in the next decade. Although job displacement is currently seen as a bleak outcome for humanity, other forces now at play will reshape our relations with robots in more positive ways. The specter of disruption driven by technological unemployment in China, for example, could conceivably be even more dramatic than that in the United States. As China has industrialized in the past two decades, significant parts of its rural population urbanized. How will China adapt to lights-out consumer electronics manufacturing?

Probably with ease, as it turns out. The Chinese population is aging dramatically, fast enough that they will soon be under significant pressure to automate their manufacturing industries. As a consequence of China's one-child policy, governmental decisions made in the late 1970s and early 1980s have now resulted in a rapidly growing elderly population. In 2050, China will have the largest number of people over 80 years old in the world. There will be 90 million elderly Chinese compared to the United States with 32 million.[53]

Europe is also aging quickly. According to European Commission data, in 2050 there will be only 2 (reduced from 4 today) people of working age in Europe for each person over 65, and an estimated 84 million people with age-related health problems.[54] The European Union views the demographic shift as a significant one and projects the emergence of a $17.6 billion market for elder-care robots in Europe by as early as 2016. The United States faces an aging scenario that is in many ways similar, although not as extreme, to Asian and European societies. Despite the fact that the United States is aging more slowly than some other countries—in part because of continuing significant immigration inflow—the "dependency ratio" will continue to rise. That means that the number of children and elderly will shift from 59 youngsters and elderly people per 100 work-age adults in 2005 to 72 per 100 in 2050.[55] The retirement of baby boomers in the United States—people who turn 65—is now taking place at the rate of roughly 10,000 each day, and that rate will continue for the next 19 years.[56]

How will the world's industrial societies care for their aging populations? An aging world will dramatically transform the conversation about robotics during the next decade from the fears about automation to new hope for augmentation. *Robot & Frank* is an amusing, thoughtful, and possibly prophetic 2012 film set in the near future, where it depicts the relationship between a retired ex-convict in the first stages of dementia and his robot caregiver. How ironic if caregiving robots like Frank's were to arrive just in time to provide a technological safety net for the world's previously displaced, now elderly population.

4 | THE RISE, FALL, AND RESURRECTION OF AI

Sitting among musty boxes in an archive at Stanford University in the fall of 2010, David Brock felt his heart stop. A detail-oriented historian specializing in the semiconductor industry, Brock was painstakingly poring over the papers of William Shockley for his research project on the life of Intel Corp. cofounder Gordon Moore. After leading the team that coinvented the transistor at Bell Labs, Shockley had moved back to Santa Clara County in 1955, founding a start-up company to make a new type of more manufacturable transistor. What had been lost, until Brock found it hidden among Shockley's papers, was a bold proposal the scientist had made in an effort to persuade Bell Labs, in 1951 the nation's premier scientific research institution, to build an "automatic trainable robot."

For decades there have been heated debates about what led to the creation of Silicon Valley, and one of the breezier explanations is that Shockley, who had grown up near downtown Palo Alto, decided to return to the region that was once the nation's fruit capital because his mother was then in ill health. He located Shockley Semiconductor Laboratory on San Antonio Road in Mountain View, just south of Palo Alto and across the freeway from where Google's sprawling corporate campus is today. Moore was one of the first employees at the fledgling transistor company and would later become a member of the "traitorous eight," the group of engineers who, because of Shockley's tyrannical management style, defected from his start-up to start a competing firm. The defection is part of the Valley's most sacred lore as an example of the intellectual and technical freedom that would make the region an entrepreneurial hotbed unlike anything the world had previously seen. Many have long believed that Shockley's decision to locate his transistor company in Mountain View was the spark that ignited Silicon Valley. However, it is more interesting to ask what Shockley was trying to accomplish. He has long been viewed as an early entrepreneur, fatally flawed as a manager. Still, his entrepreneurial passion has served as a model for generations of technologists. But that was only part of the explanation.

Brock sat in the Stanford archives staring at a yellowing single-page proposal titled the "A.T.R. Project." Shockley, true to his temper, didn't mince words: "The importance of the project described below is probably greater than any previously considered by the Bell System," he began. "The foundation of the largest industry ever to exist may well be built upon its development. It is possible that the progress achieved by industry in the next two or three decades will be directly dependent upon the vigor with which projects of this class are developed." The purpose of the project was, bluntly, "the substitution of machines for men in production." Robots were necessary because generalized automation systems lacked both the dexterity and the perception of human workers. "Such mechanization will achieve the ultimate conceivable economy on very long runs but will be impractical on short runs," he wrote.

Moreover, his original vision was not just about creating an "automatic factory," but a trainable robot that could be "readily modified to perform any one of a wide variety of operations." His machine would be composed of "hands," "sensory organs," "memory," and a "brain."[1]

Shockley's inspiration for a humanlike factory robot was that assembly work often consists of a myriad of constantly changing unique motions performed by a skilled human worker, and that such a robot was the breakthrough needed to completely replace human labor. His insight was striking because it came at the very dawn of the computer age, before the impact of the technology had been grasped by most of the pioneering engineers. At the time it was only a half decade since ENIAC, the first general purpose digital computer, had been heralded in the popular press as a "giant brain," and just two years after Norbert Wiener had written his landmark *Cybernetics,* announcing the opening of the Information Age.

Shockley's initial insight presaged the course that automation would take decades later. For example, Kiva Systems, a warehouse automation system acquired in 2012 by Amazon for $775 million, had the insight that the most difficult functions to automate in the modern warehouse were ones that required human eyes and hands, like identifying and grasping objects. Without perception and dexterity, robotic systems are limited to the most repetitive jobs, and so Kiva took the obvious intermediate step and built mobile robots that carried items to stationary human workers. Once machine perception and robotic hands became better and cheaper, humans could disappear entirely.

Indeed, Amazon made an exception to its usual policy of secrecy and invited the press to tour one of its distribution facilities in Tracy, California, during the Christmas buying season in December of 2014. What those on the press tour did not see was the development of an experimental station inside the facility where a robot arm performed the "piece pick" operations—the work now reserved for humans. Amazon is experimenting with a Danish robot arm designed to do the remaining human tasks.

In the middle of the last century, while Shockley expressed no moral qualms about using trainable robots to displace humans, Wiener saw a potential calamity. Two years after writing *Cybernetics* he wrote *The Human Use of Human Beings,* an effort to assess the consequences of a world full of increasingly intelligent machines. Despite his reservations, Wiener had been instrumental in incubating what Brock describes as an "automation movement" during the 1950s.[2] He traces the start of what would become a national obsession with automation to February 2, 1955, when Wiener and Gordon Brown, the chair of the MIT electrical engineering department, spoke to an evening panel in New York City attended by five hundred members of the MIT Alumni Association on the topic of "Automation: What is it?"

On the same night, on the other side of the country, electronics entrepreneur Arnold O. Beckman chaired a banquet honoring Shockley alongside Lee de Forest, inventor of the triode, a fundamental vacuum tube. At the event Beckman and Shockley discovered they were both "automation enthusiasts."[3] Beckman had already begun to refashion Beckman Instruments around automation in the chemical industries, and at the end of the evening Shockley agreed to send Beckman a copy of his newly issued patent for an electro-optical eye. That conversation led to Beckman funding Shockley Semiconductor Laboratory as a Beckman Instruments subsidiary, but passing on the opportunity to purchase Shockley's robotic eye. Shockley had written his proposal to replace workers with robots amid the nation's original debate over "automation," a term popularized by John Diebold in his 1952 book *Automation: The Advent of the Automatic Factory.*

Shockley's prescience was so striking that when Rodney Brooks, himself a pioneering roboticist at the Stanford Artificial Intelligence Laboratory in the 1970s, read Brock's article in *IEEE Spectrum* in 2013, he passed Shockley's original 1951 memo around his company, Rethink Robotics, and asked his employees to guess when the memo had been written. No one came close. That memo predates by more than a half century Rethink's Baxter robot, introduced in the fall of 2012. Yet Baxter is almost exactly what Shockley proposed in the 1950s—a trainable robot with an expressive "face" on an LCD screen, "hands," "sensory organs," "memory," and, of course, a "brain."

The philosophical difference between Shockley and Brooks is that Brooks's intent has been for Baxter to cooperate with human workers rather than replace them, taking over dull, repetitive tasks in a factory and leaving more creative work for humans. Shockley's original memo demonstrates that Silicon Valley had its roots in the fundamental paradox that technology both augments and dispenses with humans. Today the paradox remains sharper than ever. Those who design the systems that increasingly reshape and define the Information Age are making choices to build humans in or out of the future.

Silicon Valley's hidden history presages Google's more recent "moon shot" effort to build mobile robots. During 2013 Google quietly acquired many of the world's best roboticists in an effort to build a business claiming leadership in the next wave of automation. Like the secretive Google car project, the outlines of Google's mobile robot business have remained murky. It is still unclear whether Google as a company will end up mostly augmenting or replacing humans, but today the company is dramatically echoing Shockley's six-decade-old trainable robot ambition.

The dichotomy between AI and IA had been clear for many years to Andy Rubin, a robotics engineer who had worked for a wide range of Silicon Valley companies before coming to Google to build the company's smartphone business in 2005. In

2013 Rubin had left his post as head of the company's Android phone business and begun quietly acquiring some of the best robotics companies and technologists in the world. He found a new home for the business on California Ave., on the edge of Stanford Industrial Park just half a block away from the original Xerox PARC laboratory where the Alto, the first modern personal computer, was designed. Rubin's building was unmarked, but an imposing statue of a robot in an upstairs atrium was visible from the street below. That is, until one night the stealthy roboticists received an unhappy call from the neighbor directly across the street. The eerie-looking robot was giving their young son nightmares. The robot was moved inside where it was no longer visible to the outside world.

Years earlier, Rubin, who was also a devoted robot hobbyist, had helped fund Stanford AI researcher Sebastian Thrun's effort to build Stanley, the autonomous Volkswagen that would eventually win a $2 million DARPA prize for navigating unaided through more than a hundred miles of California desert. "Personal computers are growing legs and beginning to move around in the environment," Rubin said in 2005.[4] Since then there has been a growing wave of interest in robotics in Silicon Valley. Andy Rubin was simply an early adopter of Shockley's original insight.

However, during the half decade after Shockley's 1955 move to Palo Alto, the region became ground zero for social, political, and technological forces that would reshape American society along lines that to this day define the modern world. Palo Alto would be transformed from its roots as a sleepy college town into one of the world's wealthiest communities. However, during the 1960s and 1970s, the Vietnam War, civil rights movement, and rise of the counterculture all commingled with the arrival of microprocessors, personal computing, and computer networking.[5] In a handful of insular computer laboratories, hackers and engineers found shelter from a fractious world. By 1969, the year Richard Nixon was inaugurated president, Seymour Hersh reported the My Lai massacre, and astronauts Neil Armstrong and Buzz Aldrin walked on the moon. Americans had for the first time traveled to another world, but the nation was at the same time mired in disastrous foreign conflict.

The year 1968 had seen the premiere of the movie *2001: A Space Odyssey* painting a stark view of both the potential and pitfalls of artificial intelligence. HAL—the computer that felt impelled to violate Asimov's laws of robotics, a 1942 dictum forbidding machines to injure humans, even to ensure their survival—had defined the robot in popular culture. By the late 1960s, science-fiction writers were the nation's technology seers and AI had become a promising new technology in the form of computing and robotics—playing out both in visions of technological paradise and as populist paranoia. The future seemed almost palpable in a nation that had literally gone from *The Flintstones* to *The Jetsons* between 1960 and 1963.

Amid this cultural turmoil Charlie Rosen began building the world's first real robot as a platform for conducting artificial intelligence experiments. Rosen was a Canadian-born applied physicist who was thinking about a wide range of problems related to computing, including sensors, new kinds of semiconductors, and artificial intelligence. He was something of a renaissance man: he coauthored one of the early textbooks on transistors and had developed an early interest in neural nets—computer circuits that showed promise in recognizing patterns, "learning" by simulating behavior of biological neurons.

As a result, Stanford Research Institute became one of the two or three centers of research on neural nets and perceptrons, efforts to mimic human forms of biological learning. Rosen was a nonstop fount of ideas, continually challenging his engineers about the possibility of remarkably far-out experiments. Peter Hart, a young Stanford electrical engineer who had done research on simple pattern recognizers, remembered frequent encounters with Rosen. "Hey, Pete," Rosen would say while pressing his face up to the young scientist's, close enough that Hart could see Rosen's quivering bushy eyebrows while Rosen poked his finger into Hart's chest. "I've got an idea." That idea might be an outlandish concept for recognizing speech, involving a system to capture spoken words in a shallow tank of water about three meters long, employing underwater audio speakers and a video camera to capture the standing wave pattern created by the sound.

After describing each new project, Rosen would stare at his young protégé and shout, "What are you scared of?" He was one of the early "rainmakers" at SRI, taking regular trips to Washington, D.C., to interest the Pentagon in funding projects. It was Rosen who was instrumental in persuading the military to fund Doug Engelbart for his original idea of augmenting humans with computers. Rosen also wrote and sold the proposal to develop a mobile "automaton" as a test bed for early neural networks and other AI programs. At one meeting with some Pentagon generals he was asked if this automaton could carry a gun. "How many do you need?" was his response. "I think it should easily be able to handle two or three."

It took the researchers a while to come up with a name for the project. "We worked for a month trying to find a good name for it, ranging from Greek names to whatnot, and then one of us said, 'Hey, it shakes like hell and moves around, let's just call it Shakey,'"[6] Hart recalled.

Eventually Rosen would become a major recipient of funding from the Defense Advanced Research Projects Agency at the Pentagon, but before that he stumbled across another source of funding, also inside the military. He managed to get an audience with one of the few prominent women in the Pentagon, mathematician Ruth Davis. When Rosen told her he wanted to build an intelligent machine, she exclaimed, "You mean it could be a sentry? Could you use it to replace a soldier?" Rosen confided that he didn't think robot soldiers would be on the scene anytime soon, but he wanted to start testing prerequisite ideas about machine vision, planning, problem-solving, and understanding human language. Davis became enthused about the idea and was an early funder of the project.

Shakey was key because it was one of just a handful of major artificial intelligence projects that began in the 1960s, causing an explosion of early work in AI that would reverberate for decades. Today Shakey's original DNA can be found in everything from the Kiva warehouse robot and Google's autonomous car to Apple's Siri intelligent assistant. Not only did it serve to train an early generation of researchers, but it would be their first point of engagement with technical and moral challenges that continue to frame the limits and potential of AI and robotics today.

Many people believed Shakey was a portent for the future of AI. In November 1970 *Life* magazine hyped the machine as something far more than it actually was. The story appeared alongside a cover story about a coed college dormitory, ads for a car with four-wheel drive, and a Sony eleven-inch television. Reporter Brad Darrach's first-person account took great liberties with Shakey's capabilities in an effort to engage with the coming-of-age complexities of the machine era. He quoted a researcher at the nearby Stanford Artificial Intelligence Laboratory as acknowledging that the field had so far not been able to endow machines with complex emotional reactions such as human orgasms, but the overall theme of the piece was a reflection of the optimism that was then widespread in the robotics community.

The SRI researchers, including Rosen and his lieutenants, Peter Hart and Bert Raphael, were dismayed by a description claiming that Shakey was able to roll freely through the research laboratory's hallways at a faster-than-human-walking clip, pausing only to peer in doorways while it reasoned in a humanlike way about the world around it. According to Raphael, the description was particularly galling because the robot had not even been operational when Darrach visited. It had been taken down while it was being moved to a new control computer.[7]

Marvin Minsky, the MIT AI pioneer, was particularly galled and wrote a long rebuttal, accusing Darrach of fabricating quotes. Minsky was quoted saying that the human brain was just a computer made out of "meat." However, he was most upset at being linked to an assertion: "In from three to eight years we will have a machine with the general intelligence of an average human being. I mean a machine that will be able to read Shakespeare, grease a car, play office politics, tell a joke, have a fight. At that point the machine will begin to educate itself with fantastic speed. In a few months it will be at genius level and a few months after that its powers will be incalculable."[8] In hindsight, Darrach's alarms seem almost quaint today. Whether clueless or willfully deceptive, his broader point was simply that sooner or later—and he clearly wanted the reader to believe it was sooner—society would have to decide how they would live with their cybernetic offspring.

Indeed, despite his frustration with the inaccurate popularization of Shakey, two years later, in a paper presented at a technical computing and robotics conference in Boston, Rosen would echo Darrach's underlying theme. He rejected the idea of a completely automated, "lights-out factory" in the "next generation" largely because of the social and economic chaos that would ensue. Instead, he predicted that by the end of the 1970s, the arrival of factory and service robots (under the supervision of humans) would eliminate repetitive tasks and drudgery. The arrival of the robots would be accompanied by a new wave of technological unemployment, he argued, and it was incumbent upon society to begin rethinking issues such as the length of the workweek, retirement age, and lifetime service.[9]

For more than five years, SRI researchers attempted to design a machine that was nominally an exercise in pure artificial intelligence. Beneath the veneer of science, however, the Pentagon was funding the project with the notion that it might one day lead to a military robot capable of tracking the enemy without risking lives of U.S. or allied soldiers. Shakey was not only the touchstone for much of modern AI research as well as projects leading to the modern augmentation community—it was also the original forerunner of the military drones that now patrol the skies over Afghanistan, Iraq, Syria, and elsewhere.

Shakey exemplified the westward migration of computing and early artificial intelligence research during the 1960s. Although Douglas Engelbart, whose project was just down the hall, was a West Coast native, many others were migrants. Artificial intelligence as a field of study was originally rooted in a 1956 Dartmouth College summer workshop where John McCarthy was a young mathematics professor. McCarthy had been born in 1927 in Boston of an Irish Catholic father and Lithuanian Jewish mother, both active members of the U.S. Communist Party. His parents were intensely intellectual and his mother committed to the idea that her children could pursue any interests they chose. At twelve McCarthy encountered Eric Temple Bell's *Men of Mathematics,* a book that helped determine the career of many of the best and brightest of the era including scientists Freeman Dyson and Stanislaw Ulam. McCarthy was viewed as a high school math prodigy and only applied to Caltech, where Temple Bell was a professor, something he later decided had been an act of "arrogance." On his application he described his plans in a single sentence: "I intend to be a professor of mathematics." Bell's book had given him a realistic view of what that path would entail. McCarthy had decided that mathematicians were rewarded principally by the quality of their research, and he was taken with the idea of the self-made intellectual.

At Caltech he was an ambitious student. He jumped straight to advanced calculus and simultaneously a range of other courses including aeronautical engineering. He was drafted relatively late in the war, so his army career was more about serving as a cog in the bureaucracy than combat. Stationed close to home at Fort MacArthur in the port city of San Pedro, California, he began as a clerk, preparing discharges, then promotions for soldiers leaving the military. He made his way to Princeton for graduate school and promptly paid a visit to John von Neumann, the applied mathematician and physicist who would become instrumental in defining the basic design of the modern computer.

At this point the notion of "artificial intelligence" was fermenting in McCarthy's mind, but the coinage had not yet come to him. That wouldn't happen for another half decade in conjunction with the summer 1956 Dartmouth conference. He had first come to the concept in grad school when attending the Hixon Symposium on Cerebral Mechanisms in Behavior at Caltech.[10] At that point there weren't programmable computers, but the idea was in the air. Alan Turing, for example, had written about the possibility the previous year, to receptive audiences on both sides of the Atlantic. McCarthy was thinking about intelligence as a mathematical abstraction rather than something realizable—along the lines of Turing—through building an actual machine. It was an "automaton" notion of creating human intelligence, but not of the kind of software cellular automata that von Neumann would later pursue. McCarthy focused instead on an abstract notion of intelligence that was capable of interacting with the environment. When he told von Neumann about it, the scientist exclaimed, "Write it up!" McCarthy thought about the idea a lot but never published anything. Years later he would express regret at his inaction. Although his thesis at Princeton would focus on differential equations, he also developed an interest in logic, and a major contribution to the field of artificial intelligence would later come from his application of mathematical logic to common sense reasoning. He arrived at Princeton a year after Marvin

Minsky and discovered that they were both already thinking about the idea of artificial intelligence. At the time, however, there were no computers to allow them to work with the ideas, and so the concept would remain an abstraction.

As a graduate student, McCarthy was a contemporary of John Forbes Nash, the mathematician and Nobel laureate who would later be celebrated in Sylvia Nasar's 1998 biography, *A Beautiful Mind*. The Princeton graduate students made a habit of playing practical jokes on each other. McCarthy, for example, fell victim to a collapsing bed. He found that another graduate student was a double agent in their games, plotting with McCarthy against Nash while at the same time plotting with Nash against McCarthy. Game theory was in fashion at the time and Nash later received his Nobel Prize in economics for contributions to that field.

During the summer of 1952 both McCarthy and Minsky were hired as research assistants by mathematician and electrical engineer Claude Shannon at Bell Labs. Shannon, known as the father of "information theory," had created a simple chess-playing machine in 1950, and there was early interest in biological-growth simulating programs known as "automata," of which John Conway's 1970 Game of Life would become the most famous.

Minsky was largely distracted by his impending wedding, but McCarthy made the most of his time at Bell Labs, working with Shannon on a collection of mathematical papers that was named at Shannon's insistence *Automata Studies*.[11] Using the word "automata" was a source of frustration for McCarthy because it shifted the focus of the submitted papers away from the more concrete artificial intelligence ideas and toward more esoteric mathematics.

Four years later he settled the issue when he launched the new field that now, six decades later, is transforming the world. He backed the term "artificial intelligence" as a means of "nail[ing] the idea to the mast"[12] and focusing the Dartmouth summer project. One unintended consequence was that the term implied the idea of replacing the human mind with a machine, and that would contribute to the split between the artificial intelligence and intelligence augmentation researchers. The christening of the field, however, happened in 1956 during the event that McCarthy was instrumental in organizing: the Dartmouth Summer Research Project on Artificial Intelligence, which was underwritten with funding from the Rockefeller Foundation. As a branding exercise it would prove a momentous event. Other candidate names for the new discipline included cybernetics, automata studies, complex information processing, and machine intelligence.[13]

McCarthy wanted to avoid the term "cybernetics" because he thought of Norbert Wiener, who had coined the term, as something of a bombastic bore and he chose to avoid arguing with him. He also wanted to avoid the term "automata" because it seemed remote from the subject of intelligence. There was still another dimension inherent in the choice of the term "artificial intelligence." Many years later in a book review taking issue with the academic concept known as the "social construction of technology," McCarthy took pains to distance artificial intelligence from its human-centered roots. It wasn't about human behavior, he insisted.[14]

The Dartmouth conference proposal, he would recall years later, had made no reference to the study of human behavior, "because [he] didn't consider it relevant."[15] Artificial intelligence, he argued, was not considered human behavior except as a possible hint about performing humanlike tasks. The only Dartmouth participants who focused on the study of human behavior were Allen Newell and Herbert Simon, the Carnegie Institute researchers who had already won acclaim for ingeniously bridging the social and cognitive sciences. Years later the approach propounded by the original Dartmouth conference members would become identified with the acronym GOFAI, or "Good Old-Fashioned Artificial Intelligence," an original approach centered on achieving human-level intelligence through logic and the branch of problem-solving rules called heuristics.

IBM, by the 1950s already the world's largest computer maker, had initially been involved in the planning for the summer conference. Both McCarthy and Minsky had spent the summer of 1955 in the IBM laboratory that had developed the IBM 701, a vacuum tube mainframe computer of which only nineteen were made. In the wake of the conference, several IBM researchers did important early work on artificial intelligence research, but in 1959 the computer maker pulled the plug on its AI work. There is evidence that the giant computer maker was fearful that its machines would be linked to technologies that destroyed jobs.[16] At the time the company chief executive Thomas J. Watson Jr. was involved in national policy discussions over the role and consequences of computers in automation and did not want his company to be associated with the wholesale destruction of jobs. McCarthy would later call the act "a fit of stupidity" and a "coup."[17]

During those early years McCarthy and Minsky remained largely inseparable—Minsky's future wife even brought McCarthy along when she took Minsky home to introduce him to see her parents—even though their ideas about how to pursue AI increasingly diverged. Minsky's graduate studies had been on the creation of neural nets. As his work progressed, Minsky would increasingly place the roots of intelligence in human experience. In contrast, McCarthy looked throughout his career for formal mathematical-logical ways to model the human mind.

Yet despite their initial difficulties, early on, the field remained remarkably collegial and in the hands of researchers with privileged access to the jealously guarded room-sized computers of the era. As McCarthy recalls it, the MIT Artificial

Intelligence Laboratory came into being in 1958 after both he and Minsky had joined the university faculty. One day McCarthy met Minsky in a hallway and said to him, "I think we should have an AI project." Minsky responded that he thought that was a good idea. Just then the two men saw Jerome Wiesner, then head of the Research Laboratory on Electronics, walking toward them.

McCarthy piped up, "Marvin and I want to have an AI project."

"What do you want?" Wiesner responded.

Thinking quickly on his feet, McCarthy said, "We'd like a room, a secretary, a keypunch, and two programmers."

To which Wiesner replied, "And how about six graduate students?"

Their timing would prove to be perfect. MIT had just received a large government grant "to be excellent," but no one really knew what "excellent" meant. The grant supported six mathematics graduate students at the time, but Wiesner had no idea what they would do. So for Wiesner, McCarthy and Minsky were a serendipitous solution.[18]

The funding grant came through in the spring of 1958, immediately in the wake of the Soviet Sputnik satellite. U.S. federal research dollars were just starting to flow in large amounts to universities. It was widely believed that the generous support of science would pay off for the U.S. military, and that year President Eisenhower formed the Advanced Research Projects Agency, ARPA, to guard against future technological surprises.

The fortuitous encounter by the three men had an almost unfathomable impact on the world. A number of the "six graduate students" were connected with the MIT Model Railway Club, an unorthodox group of future engineers drawn to computing as if by a magnet. Their club ethos would lead directly to what became the "hacker culture," which held as its most prized value the free sharing of information.[19] McCarthy would help spread the hacker ethic when he left MIT in 1962 and set up a rival laboratory at Stanford University. Ultimately the original hacker culture would also foment social movements such as free/open-source software, Creative Commons, and Network Neutrality movements. While still at MIT, McCarthy, in his quest for a more efficient way to conduct artificial intelligence research, had invented computer time-sharing, as well as the Lisp programming language. He had an early notion that his AI, when it was perfected, would be interactive and logical to design on a computing system shared by multiple users, rather than requiring users to sign up to use the computer one at a time.

When MIT decided to do a survey on the wisdom of building a time-sharing system instead of immediately building what McCarthy had proposed, he decided to head west. Asking university faculty and staff what they thought of computer time-sharing would be like surveying ditchdiggers about the value of a steam shovel, he would later grouse.[20]

He was thoroughly converted to the West Coast counterculture. Although he had long since left the Communist Party, he was still on the Left and would soon be attracted to the antiestablishment community around Stanford University. He took to wearing a headband to pair with his long hair and became an active participant in the Free University that sprang up on the Midpeninsula around Stanford. Only when Russia crushed the Czech uprising in 1968 did he experience his final disillusionment with socialism. Not long afterward, while arguing over the wisdom of nonviolence during a Free U meeting, one of the radicals threatened to kill McCarthy, and he consequently ricocheted permanently to the Right. Not long afterward he registered as a Republican.

At the same time his career blossomed. Being a Stanford professor was a hunting license for funding and on his way to Stanford he turned to his friend J. C. R. Licklider, a former MIT psychologist, who headed ARPA's Information Processing

Techniques Office beginning in 1962. Licklider had collaborated with McCarthy on an early paper on time-sharing and he funded an ambitious time-sharing program at MIT after McCarthy moved to Stanford. McCarthy would later say that he never would have left if he had known that Licklider would be pushing time-sharing ideas so heavily.

On the West Coast, McCarthy found few bureaucratic barriers and quickly built an artificial intelligence lab at Stanford to rival the one at MIT. He was able to secure a computer from Digital Equipment Corporation and found space in the hills behind campus in the D.C. Power Laboratory, in a building and on land donated to Stanford by GTE after the telco canceled a plan for a research lab on the West Coast.

The Stanford Artificial Intelligence Laboratory quickly became a California haven for the same hacker sensibility that had spawned at MIT. Smart young computer hackers like Steve "Slug" Russell and Whitfield Diffie followed McCarthy west, and during the next decade and a half a startling array of hardware engineers and software designers would flow through the laboratory, which maintained its countercultural vibe even as McCarthy became politically more conservative. Both Steve Jobs and Steve Wozniak would hold on to sentimental memories of their visits as teenagers to the Stanford laboratory in the hills. SAIL would become a prism through which a stunning group of young technologists as well as full-blown industries would emerge.

Early work in machine vision and robotics began at SAIL, and the laboratory was indisputably the birthplace of speech recognition. McCarthy gave Raj Reddy his thesis topic on speech understanding, and Reddy went on to become the seminal researcher in the field. Mobile robots, paralleling Shakey at Stanford Research Institute, would be pursued at SAIL by researchers like Hans Moravec and later Rodney Brooks, both of whom became pioneering robotics researchers at Carnegie Mellon and MIT, respectively.

It proved to be the first golden era of AI, with research on natural language understanding, computer music, expert systems, and video games like Spacewar. Kenneth Colby, a psychiatrist, even worked on a refined version of Eliza, the online conversation system originally developed by Joseph Weizenbaum at MIT. Colby's simulated person was known as "Parry," with an obliquely bent paranoid personality. Reddy, who had previous computing experience using an early IBM mainframe called the 650, remembered that the company had charged $1,000 an hour for access to the machine. Now he found he "owned" a computer that was a hundred times faster for half of each day—from eight o'clock in the evening until eight the next morning. "I thought I had died and gone to heaven," he said.[21]

McCarthy's laboratory spawned an array of subfields, and one of the most powerful early on was known as knowledge engineering, pioneered by computer scientist Ed Feigenbaum. Begun in 1965, his first project, Dendral, was a highly influential early effort in the area of software expert systems intended to capture and organize human knowledge, and was initially intended to help chemists identify unknown organic molecules. It was a cooperative project among computer scientists Feigenbaum and Bruce Buchanan and two superstars from other academic fields—Joshua Lederberg, a molecular biologist, and Carl Djerassi, a chemist known for inventing the birth control pill—to automate the problem-solving strategies of an expert human organic chemist.

Buchanan would recall that Lederberg had a NASA contract related to the possibility of life on Mars and that mass spectrometry would be an essential tool in looking for such life: "That was, in fact, the whole Dendral project laid out with a very specific application, namely, to go to Mars, scoop up samples, look for evidence of organic compounds,"[22] recalled Buchanan. Indeed, the Dendral project began in 1965 in the wake of a bitter debate within NASA over what the role of humans would be in the moon mission. Whether to keep a human in the control loop was sharply debated inside the agency at the dawn of spaceflight, and is again today, decades later, concerning a manned mission to Mars.

The original AI optimism that blossomed at SAIL would hold sway throughout the sixties. It is now lost in history, but Moravec, who as a graduate student lived in SAIL's attic, recalled years later that when McCarthy first set out the original proposal he told ARPA that it would be possible to build "a fully intelligent machine" in the space of a decade.[23] From the distance of more than a half century, it seems both quixotic and endearingly naive, but from his initial curiosity in the late 1940s, before there were computers, McCarthy had defined the goal of creating machines that matched human capabilities.

Indeed, during the first decade of the field, AI optimism was immense, as was obvious from the 1956 Dartmouth workshop:

> The study is to proceed on the basis of the conjecture that every aspect of learning or any other feature of intelligence can in principle be so precisely described that a machine can be made to simulate it. An attempt will be made to find how to make machines use language, form abstractions and concepts, solve kinds of problems now reserved for humans, and improve themselves. We think that a significant advance can be made in one or more of these problems if a carefully selected group of scientists work on it together for a summer.[24]

Not long afterward Minsky would echo McCarthy's optimism, turning a lone graduate student loose on the problem of machine vision, figuring that it was a suitable problem to be solved as a summer project.[25] "Our ultimate objective is to make programs that learn from their experience as effectively as humans do," McCarthy wrote.[26]

As part of that effort he created a laboratory that was a

paradise for researchers who wanted to mimic humans in machine form. At the same time it would also create a cultural chasm that resulted in a computing world with two separate research communities—those who worked to replace the human and those who wanted to use the same technologies to augment the human mind. As a consequence, for the past half century an underlying tension between artificial intelligence and intelligence augmentation—AI versus IA—has been at the heart of progress in computing science as the field has produced a series of ever more powerful technologies that are transforming the world.

It is easy to argue that AI and IA are simply two sides of the same coin. There is a fundamental distinction, however, between approaches to designing technology to benefit humans and designing technology as an end in itself. Today, that distinction is expressed in whether increasingly capable computers, software, and robots are designed to assist human users or to replace them. Early on some of the researchers who passed through SAIL rebelled against McCarthy-style AI. Alan Kay, who pioneered the concept of the modern personal computer at Xerox during the 1970s, spent a year at SAIL, and would later say it was one of the least productive years of his career. He already had fashioned his Dynabook idea—"a personal computer for children of all ages"[27]—that would serve as the spark for a generation of computing, but he remained an outsider in the SAIL hacker culture. For others at SAIL, however, the vision was clear: machines would soon match and even replace humans. They were the coolest things around and in the future they would meet and then exceed the capabilities of their human designers.

You must drive several miles from the Carnegie Mellon University campus to reach a pleasantly obscure Pittsburgh residential neighborhood to find Hans Moravec. His office is tucked away in a tiny apartment at the top of a flight of stairs around the corner from a small shopping street. Inside, Moravec, who retains his childhood Austrian accent, has converted a two-room apartment into a hideaway office where he can concentrate without interruption. The apartment opens into a cramped sitting room housing a small refrigerator. At the back is an even smaller office, with curtains down, dominated by large computer displays.

Several decades ago, when he captured the public's attention as one of the world's best-known robot designers, magazines often described him as "robotic." In person, he is anything but, breaking out in laughter frequently and with a self-deprecating sense of humor. Still an adjunct professor at the Robotics Institute at Carnegie Mellon, where he taught for many years, Moravec, one of John McCarthy's best-known graduate students, has largely vanished from the world he helped create.

When Robert M. Geraci, a religious studies professor at Manhattan College and author of *Apocalyptic AI: Visions of Heaven in Robotics, Artificial Intelligence, and Virtual Reality* (2010), came to Pittsburgh to conduct his research several years ago, Moravec politely declined to see him, citing his work on a recent start-up. Geraci is one of a number of authors who have painted Moravec as the intellectual cofounder, with Ray Kurzweil, of a techno-religious movement that argues that humanity will inevitably be subsumed as a species by the AIs and robots we are now creating. In 2014 this movement gained generous exposure as high-profile technological and scientific luminaries such as Elon Musk and Stephen Hawking issued tersely worded warnings about the potential threat that futuristic AI systems hold for the human species.

Geraci's argument is that there is a generation of computer technologists who, in looking forward to the consequences of their inventions, have not escaped Western society's religious roots but rather recapitulated them. "Ultimately, the promises of Apocalyptic AI are almost identical to those of Jewish and Christian apocalyptic traditions. Should they come true, the world will be, once again, a place of magic,"[28] Geraci wrote. For the professor of religion, the movement could in fact be reduced to the concept of alienation, which in his framing is mainly about the overriding human fear of dying.

Geraci's conception of alienation isn't simply a 1950s James Dean–like disconnect from society. Yet it is just as hard to pin Moravec on the more abstract concept of fear of death. The robotics pioneer became legendary for taking up residence in the attic of McCarthy's SAIL lab during the 1970s, when it was a perfect counterculture world for the first generation of computer hackers who discovered that the machines they had privileged access to could be used as "fantasy amplifiers."

During the 1970s, McCarthy continued to believe that artificial intelligence was within reach even with the meager computing resources then at hand, famously noting that a working AI would require: "1.8 Einsteins and one-tenth the resources of the Manhattan Project."[29] In contrast, Moravec's perspective was rooted in the rapidly accelerating evolution of computing technology. He quickly grasped the implications of Moore's law—the assertion that over time computing power would increase exponentially—and extended that observation to what he believed would be the logical conclusion: machine intelligence was inevitable and moreover it would happen relatively soon. He summed up the obstacles faced by the AI field in the late 1970s succinctly:

> The most difficult tasks to automate, for which computer performance to date has been most disappointing, are those that humans do most naturally, such as seeing, hearing and common sense reasoning. A major reason for the difficulty has become very clear to me in the course of my work on computer vision. It is simply that the machines with which we are working are still a hundred

thousand to a million times too slow to match the performance of human nervous systems in those functions for which humans are specially wired. This enormous discrepancy is distorting our work, creating problems where there are none, making others impossibly difficult, and generally causing effort to be misdirected.[30]

He first outlined his disagreement with McCarthy in 1975 in the SAIL report "The Role of Raw Power in Intelligence."[31] It was a powerful manifesto that steeled his faith in the exponential increase in processing power and simultaneously convinced him that the current limits were merely a temporary state of affairs. The lesson he drew early on, and to which he would return throughout his career, was that if you were stymied as an AI designer, just wait a decade and your problems would be solved by the inexorable increase in computing performance. In a 1978 essay for the science-fiction magazine *Analog,* he laid out his argument for a wider public. Indeed in the *Analog* essay he still retained much of McCarthy's original faith that machines would cross the level of human intelligence in about a decade: "Suppose my projections are correct, and the hardware requirements for human equivalence are available in 10 years for about the current price of a medium large computer," he asked. "What then?"[32] The answer was obvious. Humans would be "outclassed" by the new species we were helping to evolve.

After leaving Stanford in 1980, Moravec would go on to write two popular books sketching out the coming age of intelligent machines. *Mind Children: The Future of Robot and Human Intelligence* (1988) contains an early detailed argument that the robots that he has loved since childhood are in the process of evolving into an independent intelligent species. A decade later he refined the argument in *Robot: Mere Machine to Transcendent Mind* (1998).

Significantly, although it is not widely known, Doug Engelbart had made the same observation, that computers would increase in power exponentially, at the dawn of the interactive computing age in 1960.[33] He used this insight to launch the SRI-based augmentation research project that would help lead ultimately to both personal computing and the Internet. In contrast, Moravec built on his lifelong romance with robots. Though he has tempered his optimism, his overall faith never wavered. During the 1990s, in addition to writing his second book, he took two sabbaticals in an effort to hurry the process of perfecting the ability to permit machines to see and understand their environments so they could navigate and move freely.

The first sabbatical he spent in Cambridge, Massachusetts, at Danny Hillis's Thinking Machines Corporation, where Moravec hoped to take advantage of a supercomputer. But the new supercomputer, the CM-5, wasn't ready. So he contented himself with refining his code on a workstation while waiting for the machine. By the end of his stay, he realized that he only needed to wait for the power of a supercomputer to come to his desktop rather than struggle to restructure his code so it would run on a special-purpose machine. A half decade later, on a second sabbatical at a Mercedes-Benz research lab in Berlin, he again had the same realization.

Moravec still wasn't quite willing to give up and so after coming back from Germany he took a DARPA contract to continue work on autonomous mobile robotic software. But after writing two best-selling books over a decade arguing for a technological promised land, he decided it was really time to settle down and do something about it. The idea that the exponential increase of computing power would inevitably lead to artificially intelligent machines was becoming more deeply ingrained in Silicon Valley, and a slick packaging of the underlying argument was delivered in 2005 by Ray Kurzweil's *The Singularity Is Near.* "It was becoming a spectacle and it was interfering with real work," he decided. By now he had taken to heart Alan Kay's dictum that "the best way to predict the future is to invent it."

His computer cave is miles from the offices of Seegrid, the robotic forklift company he founded in 2003, but within walking distance of his Pittsburgh home. For the past decade he has given up his role as futurist and became a hermit. In a way, it is the continuation of the project he originally began as a child. Growing up in Canada, at age ten Moravec had built his first robot from tin cans, batteries, lights, and a motor. Later, in high school, he went on to build a robotic turtle capable of following a light and a robotic hand. At Stanford, he became the force behind the Stanford Cart project, a mobile robot with a TV camera that could negotiate obstacle courses. He had inherited the Cart system when he arrived at Stanford in 1971 and then gradually rebuilt the entire system.

Shakey was the first autonomous robot, but the Stanford Cart, with a long and colorful history of its own, is the true predecessor of the autonomous car. It had first come to life as a NASA-funded project in the mechanical engineering department in 1960, based on the idea that someday a vehicle would be remotely driven on the surface of the moon. The challenge was how to control such a vehicle given the 2.7-second propagation delay defining the round-trip radio signal between the Earth and the moon.

Funding for the initial project was rejected because the logic of keeping a human in the loop had won out. When in 1962 President Kennedy committed the nation to the manned exploration of the moon, the original Cart was shelved[34] as unnecessary. The robot, about the size of a card table with four bicycle wheels, sat unused until in 1966 SAIL's deputy director Les Earnest rediscovered it. He persuaded the mechanical engineering department to lend it to SAIL to experiment in making an autonomous vehicle. Eventually, using the computing power of the SAIL mainframe, a graduate student was able to program the robot to follow a white line on the floor at a speed of less than one mile per hour. A radio control link enabled remote operation. Tracking would have been simpler with two photocell sensors, but a video camera connected to a computer was seen as a tour de force at the time.

Moravec would modify and hack the system for a decade so that ultimately it would be able to make it across a room, correctly navigating an obstacle course about half the time. The Cart failed in many ways. Attempting to simultaneously map and locate using only single camera data, Moravec had undertaken one of the hardest problems in AI. His goal was to build an accurate three-dimensional model of the world as a key step toward understanding it.

At the time the only feedback came from seeing how far the Cart had moved. It didn't have true stereoscopic vision, so the Cart lacked depth perception. As a cost-saving measure, he would move the camera back and forth along a bar at right angles to the field of view, making it possible for the software to calculate a stereo view from a single camera. It was an early predecessor of the software approach taken decades later by the Israeli computer vision company Mobileye.

Driving automatically was rather slow and boring, and so with its remote link and video camera connection Moravec enjoyed controlling the Cart remotely from his computer workstation. It all seemed very futuristic to pretend that he was at the controls of a lunar rover, wandering around SAIL, which was housed in a circular building in the hills west of Stanford. Before long the driveway leading up to the lab was sporting a yellow traffic sign that read CAUTION ROBOTIC VEHICLE. The Cart did venture outside but not very successfully. Indeed it seemed to have a propensity to find trouble. One setback occurred in October of 1973 when the Cart, being driven manually, ran off an exit ramp, tipped over, leaked acid from a battery, and in the process destroyed precious electronic circuitry.[35] It took almost a year to rebuild.

Moravec would often try to drive the Cart around the building, but toward the rear of the lab the road dipped, causing the radio signal to weaken and making it hard to see exactly where the Cart was. Once, while the Cart was circling the building, he misjudged its location and made a wrong turn. Rather than returning in a circle, the robot headed down the driveway to busy Arastradero Road, which runs through the Palo Alto foothills. Moravec kept waiting for the signal from the robot to improve, but it stayed hazy. The television image was filled with static. Then to his surprise, he saw a car drive by the robot. That seemed odd. Finally he got up from his computer terminal and went outside to track the robot down physically. He walked to where he assumed the robot would be, but found nothing. He decided that someone was playing a joke on him. Finally, as he continued to hunt for the errant machine it came rolling back up the driveway with a technician sitting on it. The Stanford Cart had managed to make its way far down Arastradero Road and was driving away from the lab by the time it was captured. Since those baby steps, engineers have made remarkable progress in designing self-driving cars. Moravec's original premise that it is only necessary to wait for computing to fall in cost and grow more powerful has largely proven true.

He has quietly continued to pursue machine vision technology, but there have been setbacks. In October of 2014, his startup factory vision system declared bankruptcy and underwent a court-ordered restructuring. Despite the disappointments, only the timeline in his agenda has changed. To the question of whether the current wave of artificial intelligence and robotics will replace human labor, he responds with a twinkle that what he is about is replacing *humanity*—"labor is such a minimalist goal."

He originally sketched out his vision of the near future in his second book, *Robot: Mere Machine to Transcendent Mind*. Here, Moravec concludes there is no need to replace capitalism because it is worthwhile for evolving machines to compete against each other. "The suggestion," he said, "is in fact that we engineer a rather modest retirement for ourselves." The specter of the "end of labor," which today is viewed with growing alarm by many technologists, is a relatively minor annoyance in Moravec's worldview. Humans are adept at entertaining each other. Like many of his Singularitarian brethren, he instead wonders what we will do with a superabundance of all of society's goods and services. Democracy, he suggests, provides a path to sharing the vast accumulation of capital that will increasingly come from superproductive corporations. It would be possible, for example, to increase Social Security payments and lower the retirement age until it eventually equals birth.

In Moravec's worldview, augmentation is an interim stage of technology development, only necessary during the brief period when humans can still do things that the machines can't. Like Licklider he assumes that machines will continue to improve at a faster and faster rate, while humans will evolve only incrementally. Not by 2020—and at one point he believed 2010—but sometime reasonably soon thereafter so-called universal robots will arrive that will be capable of a wide set of basic applications. It was an idea he first proposed in 1991, and only the timeline has been altered. At some point these machines will improve to the point where they will be able to learn from experience and gradually adapt to their surroundings. He still retains his faith in Asimov's three laws of robotics. The market will ensure that robots behave humanely—robots that cause too many deaths simply won't sell very well. And at some point, in Moravec's view, machine consciousness will emerge as well.

Also in *Robot: Mere Machine to Transcendent Mind,* he argues that strict laws be passed and applied to fully automated corporations. The laws would limit the growth of these corporations—and the robots they control—and prohibit them from taking too much power. If they grow too large, an automatic antitrust mechanism will go into effect, forcing them to divide. In Moravec's future world, rogue corporations will be held in check by a society of AI-based corporations, working to protect the common good. There is nothing romantic in his worldview: "We can't get too sentimental about the robots, because unlike human beings the robots don't have this evolutionary history where their own survival is really the most important thing," he said. He still holds to the basic premise—the arrival of the AI-based corporation and the universal robot will mark a utopia that will satisfy every human desire.

His worldview isn't entirely utopian, however. There is also a darker framing to his AI/robot future. Robots will be expanding into the solar system, mining the asteroids and reproducing and building copies of themselves. This is where his ideas begin to resemble *Blade Runner*—the dystopian Ridley Scott movie in which androids have begun to colonize the solar system. "Something can go wrong, you will have rogue robots out there," he said. "After a while you will end up with an asteroid belt and beyond that is full of wildlife that won't have the mind-numbing restrictions that the tame robots on Earth have." Will we still need a planetary defense system to protect us from our progeny? Probably not, he reasons. This new technological life-form will be more interested in expanding into the universe—hopefully.

From his cozy and solitary command center in suburban Pittsburgh in a room full of computer screens it is easy to buy into Moravec's science-fiction vision. So far, however, there is precious little solid evidence that there will be a rapid technological acceleration that will bring about the AI promised land in his lifetime. Despite the reality that we don't yet have self-driving cars and the fact that he has been forced to revise the timing of his estimates, he displays the curves on his giant computer screens and remains firm in his basic belief that society is still on track to create its successor species.

Will humans join in this grand adventure? Although he proposed the idea of uploading a human mind into a computer in *Mind Children,* Moravec is not committed to Ray Kurzweil's goal of "living long enough to live forever." Kurzweil is undergoing extraordinary and questionable medical procedures to extend his life. Moravec confines his efforts to survive until 2050 to eating well and walking frequently, and at the age of sixty-four, he does perhaps stand a plausible chance of surviving that long.

During the 1970s and 1980s the allure of artificial intelligence would draw a generation of brilliant engineers, but it would also disappoint. When AI failed to deliver on its promise they would frequently turn to the contrasting ideal of intelligence augmentation.

Sheldon Breiner grew up in a middle-class Jewish family in St. Louis and, from an early age, was extraordinarily curious about virtually everything he came in contact with. He chose college at Stanford during the 1950s, in part to put as much distance as possible between himself and the family bakery. He wanted to see the world, and even in high school realized that if he stayed in St. Louis his father would likely compel him to take over the family business.

After graduating he traveled in Europe, spent some time in the army reserve, and then came back to Stanford to become a geophysicist. Early on he had become obsessed with the idea that magnetic forces might play a role in either causing or perhaps predicting earthquakes. In 1962 he had taken a job at Varian Associates, an early Silicon Valley firm making a range of magnetometers. His assignment was to find new applications for these instruments that could detect minute variations in the Earth's magnetic field. Varian was the perfect match for Breiner's 360-degree intelligence. For the first time highly sensitive magnetometers were becoming portable, and there was a willing market for clever new applications that would range from finding oil to airport security. Years later Breiner would become something of a high-tech Indiana Jones, using the technology to explore archaeological settings. In Breiner's expert hands, Varian magnetometers would find avalanche victims, buried treasure, missing nuclear submarines, and even buried cities. Early on he conducted a field experiment from a site behind Stanford, where he measured the electromagnetic pulse (EMP) generated by a 1.4-megaton nuclear detonation 250 miles above the Earth. The classified test, known as Starfish Prime, led to new understanding about the impact of nuclear explosions on Earth-based electronics.

For his 1967 doctoral thesis he set out to explore the question of whether minute changes in the huge magnetic forces deep in the Earth could play a role in predicting earthquakes. He set up an array of magnetometers in individual trailers along a 120-mile stretch on the San Andreas Fault and used phone lines to send the data back to a laboratory in an old shack on the Stanford campus. There he installed a pen plotter that would record signals from the various magnetometers. It was an ungainly device that pushed rather than pulled a roll of chart paper under five colored ink pens. He hired a teenager from a local high school to change the paper and time-stamp the charts, but the device was so flawed that it caused the paper to ball up in huge piles almost every other day. He redesigned the system around a new digital printer from Hewlett-Packard and the high school student, who had been changing the paper for a dollar a day, was an early automation casualty.

Later, Breiner was hired by Hughes Corp. to work on the design of a deep ocean magnetometer to be towed by the *Glomar Explorer*. The cover story was to hunt for minerals such as manganese nodules on the seabed at depths of ten thousand to twelve thousand feet. A decade later the story leaked that the actual mission was a Central Intelligence Agency operation to find and raise a sunken Soviet submarine from the bottom of the Pacific Ocean. In 1968, after the assassination of Robert Kennedy, Breiner was asked by the White House science advisor to demonstrate technology to detect hidden weapons. He went to the Executive Office Building and demonstrated a relatively simple scheme employing four magnetometers that would become the basis for the modern metal detectors still widely used in airports and public buildings.[36]

Ultimately Breiner was able to demonstrate evidence of magnetic variation along the fault associated with earthquakes, but the data was clouded by geomagnetic activity and his hypothesis did not gain wide acceptance. He didn't let the lack of scientific certainty hold him back. At Varian he had been paid to think up a wide range of commercial applications for magnetometers and in 1969 he and five Varian colleagues founded Geometrics, a company that used airborne magnetometers to prospect for oil deposits.

He would run his oil prospecting company for seven years before selling to Edgerton, Germeshausen, and Grier (EG&G), and then work for seven more years at their subsidiary before leaving in 1983. By then, the artificial intelligence technology that had been pioneered in John McCarthy's SAIL and in the work that Feigenbaum and Lederberg were doing to capture and bottle human expertise was beginning to leak out into the surrounding environment in Silicon Valley. A *Businessweek* cover story in July of 1984 enthused, "Artificial Intelligence— It's Here!" Two months later on CBS *Evening News* Dan Rather gave glowing coverage to the SRI work in developing expert systems to hunt for mineral deposits. Bathed in the enthusiasm, Breiner would become part of a wave of technology-oriented entrepreneurs who came to believe that the time was ripe to commercialize the field.

The earlier work on Dendral in 1977 had led to a cascade of similar systems. Mycin, also produced at Stanford, was based on an "inference engine" that did if/then–style logic and a "knowledge base" of roughly six hundred rules to reason about blood infections. At the University of Pittsburgh during the 1970s a program called Internist-I was another early effort to tackle the challenge of disease diagnosis and therapy. In 1977 at SRI, Peter Hart, who began his career in artificial intelligence working on Shakey the robot, and Richard Duda, another pioneering artificial intelligence researcher, built Prospector to aid in the discovery of mineral deposits. That work would eventually get CBS's overheated attention. In the midst of all of this, in 1982, Japan announced its Fifth Generation Computer program. Heavily focused on artificial intelligence, it added an air of competition and inevitability to the AI boom that would lead to a market in which newly minted Ph.D.s could command unheard-of $30,000 annual salaries right out of school.

The genie was definitely out of the bottle. Developing expert systems was becoming a discipline called "knowledge engineering"—the idea was that you could package the expertise of a scientist, an engineer, or a manager and apply it to the data of an enterprise. The computer would effectively become an oracle. In principle that technology could be used to augment a human, but software enterprises in the 1980s would sell it into corporations based on the promise of cost savings. As a productivity tool its purpose was as often as not to displace workers.

Breiner looked around for industries where it might be easy to package the knowledge of human experts and quickly settled on commercial lending and insurance underwriting. At the time there was no widespread alarm about automation and he didn't see the problem framed in those terms. The computing world was broken down into increasingly inexpensive personal computers and more costly "workstations," generally souped-up machines for computer-aided design applications. Two companies, Symbolics and Lisp Machines, Inc., spun directly out of the MIT AI Lab to focus on specialized computers running the Lisp programming language, designed for building AI applications.

Breiner founded his own start-up, Syntelligence. Along with Teknowledge and Intellicorp, it would become one of the three high-profile artificial intelligence companies in Silicon Valley in the 1980s. He went shopping for artificial intelligence talent and hired Hart and Duda from SRI. The company created its own programming language, Syntel, which ran on an advanced workstation used by the company's software engineers. It also built two programs, Underwriting Advisor and Lending Advisor, which were intended for use on IBM PCs. He positioned the company as an information utility rather than as an artificial intelligence software publisher. "In every organization there is usually one person who is really good, who everybody calls for advice," he told a *New York Times* reporter writing about the emergence of commercial expert systems. "He is usually promoted, so that he does not use his expertise anymore. We are trying to protect that expertise if that person quits, dies or retires and to disseminate it to a lot of other people." The article, about the ability to codify human reasoning, ran on the paper's front page in 1984.[37]

When marketing his loan expert and insurance expert software packages, Breiner demonstrated dramatic, continuing cost savings for customers. The idea of automating human expertise was compelling enough that he was able to secure preorders from banks and insurance companies and investments from venture capital firms. AIG, St. Paul, and Fireman's Fund as well as Wells Fargo and Wachovia advanced $6 million for the software. Breiner stuck with the project for almost a half decade, ultimately growing the company to more than a hundred employees and pushing revenues to $10 million annually. The problem was that wasn't fast enough for his investors. In 1983 the five-year projections had been to be at $50 million of annual revenue. When the commercial market for artificial intelligence software failed to materialize quickly enough, inside the company he struggled, most bitterly with board member Pierre Lamond, a venture capitalist who was a veteran of the semiconductor industry with no software experience. Ultimately Breiner lost his battle and Lamond brought in an outside corporate manager who moved the company head-quarters to Texas, where the manager lived.

Syntelligence itself would confront directly what would be become known as the "AI Winter." One by one the artificial intelligence firms of the early 1980s were eclipsed either because they failed financially or because they returned to their roots as experimental efforts or consulting companies. The market failure became an enduring narrative that came to define artificial intelligence, with a repeated cycle of hype and failure fueled by overly ambitious scientific claims that are inevitably followed by performance and market disappointments. A generation of true believers, steeped in the technocratic and optimistic artificial intelligence literature of the 1960s, clearly played an early part in the collapse. Since then the same boom-and-bust cycle has continued for decades, even as AI has advanced.[38] Today the cycle is likely to repeat itself again as a new wave of artificial intelligence technologies is being heralded by some as being on the cusp of offering "thinking machines."

The first AI Winter had actually come a decade earlier in Europe. Sir Michael James Lighthill, a British applied mathematician, led a study in 1973 that excoriated the field for not delivering on the promises and predictions, such as the early SAIL prediction of a working artificial intelligence in a decade. Although it had little impact in the United States, the Lighthill report, "Artificial Intelligence: A General Survey," led to the curtailment of funding in England and a dispersal of British researchers from the field. In a footnote of the report the BBC arranged a televised debate on the future of AI where the targets of Lighthill's criticism were given a forum to respond. John McCarthy was flown in for the event but was unable to offer a convincing defense of his field.

A decade later a second AI Winter would descend in the United States, beginning in 1984, when Breiner managed to push Syntelligence sales to $10 million before departing. There had been warnings of "irrational exuberance" for several years when Roger Schank and Marvin Minsky raised the issue early on at a technical conference, claiming that emerging commercial expert systems contained no significant technical advances from work that had begun two decades earlier.[39] The year 1984 was also when Doug Engelbart's and Alan Kay's augmentation ideas dramatically came within the reach of every office worker. Needing a marketing analogy to frame the value of the personal computer with the launch of the Macintosh, Steve Jobs hit on the perfect metaphor for the PC. It was a "bicycle for our minds."

Pushed out of the company he had founded, Breiner went on to his next venture, a start-up company designing software for Apple's Macintosh. From the 1970s through the 1980s it was a path followed by many of Silicon Valley's best and brightest.

Beginning in the 1960s, the work that had been conducted quietly at the MIT and Stanford artificial intelligence laboratories and at the Stanford Research Institute began to trickle out into the rest of the world. The popular worldview of robotics and artificial intelligence had originally been given form by literary works—the mythology of the Prague Golem, Mary Shelley's *Frankenstein,* and Karel Čapek's pathbreaking *R. U. R. (Rossum's Universal Robots)*—all posing fundamental questions about the impact of robotics on humans life. However, as America prepared to send humans to the moon, a wave of technology-rich and generally optimistic science fiction appeared from writers like Isaac Asimov, Robert Heinlein, and Arthur C. Clarke. HAL, the run-amok sentient computer in Clarke's *2001: A Space Odyssey,* not only had a deep impact on popular culture, it changed people's lives. Even before he began as a graduate student in computer science at the University of Pennsylvania, Jerry Kaplan knew what he planned to do. The film version of *2001* was released in the spring of 1968, and over the summer Kaplan watched it six times. With two of his friends he went back again and again and again. One of his friends said, "I'm going to make movies." And he did—he became a Hollywood director. The other friend became a dentist, and Kaplan went into AI.

"I'm going to build that," he told his friends, referring to HAL. Like Breiner, he would become instrumental as part of the first generation to attempt to commercialize AI, and also like Breiner, when that effort ran aground in the AI Winter, he would turn to technologies that augmented humans instead.

As a graduate student Kaplan had read Terry Winograd's SHRDLU tour de force on interacting with computers via natural language. It gave him a hint about what was possible in the world of AI as well as a path toward making it happen. Like many aspiring computer scientists at the time, he would focus on understanding natural language. A math whiz, he was one of a new breed of computer nerds who weren't just pocket-protector-clad geeks, but who had a much broader sense of the world.

After he graduated with a degree in the philosophy of science from the University of Chicago, he followed a girlfriend to Philadelphia. An uncle hired him to work in the warehouse of his wholesale pharmaceuticals business while grooming him to one day take over the enterprise. Dismayed by the claustrophobic family business, he soon desperately needed to do something different and he remembered both a programming class he had taken at Chicago and his obsession with *A Space Odyssey*. He enrolled as a graduate student in computer science at the University of Pennsylvania. Once there he studied with Aravind Krishna Joshi, an early specialist in computational linguistics. Even though he had come in with a liberal arts background he quickly became a star. He went through the program in five years, getting perfect scores in all of his classes and writing his graduate thesis on the subject of building natural language front ends to databases.

As a newly minted Ph.D., Kaplan gave job audition lectures at Stanford and MIT, visited SRI, and spent an entire week being interviewed at Bell Labs. Both the telecommunications and computer industry were hungry for computer science Ph.D.s and on his first visit to Bell Labs he was informed that the prestigious lab had a target of hiring 250 Ph.D.s, and had no intention of hiring below average. Kaplan couldn't help pointing out that 250 was more than the entire number of Ph.D.s that the United States would produce that year. He picked Stanford, after Ed Feigenbaum had recruited him as a research associate in the Knowledge Engineering Laboratory. Stanford was not as intellectually rigorous as Penn, but it was a technological paradise. Silicon Valley had already been named, the semiconductor industry was under assault from Japan, and Apple Computer was the nation's fastest-growing company.

There was free food at corporate and academic events every evening and no shortage of "womanizing" opportunities. He bought a home in Los Trancos Woods several miles from Stanford, near SAIL, which was just in the process of moving from the foothills down to a new home on the central Stanford campus.

When he arrived at Stanford in 1979 the first golden age of AI was in full swing—graduate students like Douglas Hofstadter, the author of *Gödel, Escher, Bach: An Eternal Golden Braid;* Rodney Brooks; and David Shaw, who would later take AI techniques and transform them into a multibillion-dollar hedge fund on Wall Street, were all still around. The commercial forces that would lead to the first wave of AI companies like Intellicorp, Syntelligence, and Teknowledge were now taking shape. While Penn had been like an ivory castle, the walls between academia and the commercial world were coming down at Stanford. There was wheeling and dealing and start-up fever everywhere. Kaplan's officemate, Curt Widdoes, would soon take the software used to build the S1 supercomputer with him to cofound Valid Logic Systems, an early electronic design automation company. They used newly developed Stanford University Network (SUN) workstations. Graduate student Andy Bechtolsheim—sitting in the next room—had designed the original SUN hardware, and would soon cofound Sun Microsystems, thus commercializing the hardware he had developed as a graduate student.

Kaplan rapidly became a "biz-dev" guy. It was in the air. He had an evening consulting gig developing the software for what would become Synergy, the first all-digital music keyboard synthesizer. It was chock-full of features that have become standard on modern synthesizers, and was used to produce the soundtrack for the movie *Tron*. Like everyone at Stanford, he was making money on the side. They were all starting companies. There was a guy in the basement, Leonard Bosack, who was trying to figure out how to interconnect computers and would eventually found Cisco Systems with his wife, Sandy Lerner, to make the first network routers.

Kaplan had a research associate job at Stanford, which was great. It was equivalent to a non–tenure track teaching position, but without the pain of having to teach. There was, however, a downside. Research staff were second-class citizens to academic faculty. He was treated like the hired help, even though he could write code and do serious technical work. His role was like Scotty, the reliable engineer on the starship *Enterprise* in *Star Trek*. He was the person who made things work. Fueled in part by the Reagan-era Strategic Defense Initiative, vast new investments were being made in artificial intelligence. It was military-led spending, but it wasn't entirely about military applications. Corporate America was toying with the idea of expert systems. Ultimately the boom would lead to forty start-up companies and U.S. sales of AI-related hardware and software of $425 million in 1986. As an academic, Kaplan lasted just two years at Stanford. He received two offers to join start-ups at the same time, both in the AI world. Ed Feigenbaum, who had decided that the Stanford computer scientists should get paid for what they were already doing academically, was assembling one of the start-ups, Teknowledge. The new company would rapidly become the Cadillac of expert system consulting, also developing custom products. The other start-up was called Symantec. Decades later it would become a giant computer security firm, but at the outset Symantec began with an AI database program that overlapped with Kaplan's technical expertise.

It was a time when Kaplan seemed to have an unlimited capacity to work. He wasn't a big partier, he didn't like being interrupted, and he viewed holidays as a time to get even more accomplished. Gary Hendrix, a respected natural language researcher at SRI, approached him to help with the programming of an early demo version of a program called Q&A, the first natural language database. The idea was that unskilled users would be able to retrieve information by posing queries in normal sentences. There was no money, only a promise of stock if the project took off.

Kaplan's expertise was on natural language front ends that would allow typed questions to an expert system. What Hendrix needed, however, was a simple database back end for his demonstration. And so over a Christmas holiday at the end of 1980, Kaplan sat down and programmed one. The entire thing initially ran on an Apple II. He did it on a contingent basis and in fact he didn't get rich. The first Symantec never went anywhere commercially and the venture capitalists did a "cram down," a financial maneuver in which company founders often see their equity lose value in exchange for new investments. As a result, what little stock Kaplan owned was now worthless.

In the end he left Stanford and joined Teknowledge because he admired Lee Hecht, the University of Chicago physicist and business school professor who had been brought in to be CEO and provide adult supervision for the twenty Stanford AI refugees who were the Teknowledge shock troops. "Our founders have build [sic] more expert systems than anyone else," Hecht told *Popular Science* in 1982.[40] Teknowledge set up shop at the foot of University Ave., just off the Stanford campus, but soon moved to flashier quarters farther down the street in the one high-rise in downtown Palo Alto. In the early 1980s the office had a sleek modernist style that leaned heavily toward black.

The state-of-the-art office offered a clear indication that the new AI programs wouldn't be cheap. Just one rule for one of the expert systems would require an interviewer to spend an hour with a human expert, and a working expert system would consist of five hundred rules or more. A complete system might cost as much as $4 million to build, but Hecht, like Breiner, believed that by bottling human expertise, corporations could reap vast savings over time. A complete system might save a manufacturer as much as $100 million annually, he told the magazine. An oil company expert system that they were prototyping might save as much as $1,000 per well per day, Hecht claimed. In the article Feigenbaum also asserted that the bottleneck would be broken when computers themselves began automatically interviewing experts.[41] Hecht saw more than a hacker in Kaplan and made him a promise—if he came to Teknowledge he would teach him how to run a business. He jumped at the chance. His office was adjacent to Hecht's and he set out to build a next-generation consulting firm whose mission was to replace the labor of human experts with software.

However, in the beginning Kaplan knew nothing about the art of selling high-technology services. He was put in charge of marketing and the first thing he did was prepare a brochure describing the firm's services. From an entirely academic background he put together a trifold marketing flyer that was intended to attract corporate customers to a series of seminars on how to build an expert system featuring Feigenbaum as the star speaker. He sent out five thousand brochures. You were supposed to get a 2 percent response rate. Instead of a hundred responses, they got just three, and one was from a guy who thought they were teaching about artificial insemination. It was a rude shock for a group of AI researchers, confident that they were about to change the world overnight, that outside of the university nobody had heard of artificial intelligence. Eventually, they were able to pull together a small group of mostly large and defense-oriented corporations, making it possible for Hecht to say that there had "been inquiries from more than 50 major companies from all over the world," and Teknowledge was able to do $1 million in business in two months at the beginning of 1982.[42]

It was indeed a Cadillac operation. They wrote the programs in Lisp on fancy $20,000 Xerox Star workstations. Worse, the whole operation was buttressed by just a handful of marketers led by Kaplan. The Teknowledge worldview was, "We're smart, we're great, people should just give us money." It was completely backward, and besides, the technology didn't really work. Despite the early stumbles, however, they eventually attracted attention. One day the king of Sweden even came to visit. True to protocol his arrival had all the trappings of a regal entourage. The Secret Service showed up first to inspect the office, including the bathroom. The assembled advance team appeared to be tracking the king in real time as they waited. Kaplan was standing breathlessly at the door when a small, nondescript gentleman in standard Silicon Valley attire—business casual—walked in unaccompanied and innocently said to the young Teknowledge executive, "Where should I sit?" Flustered, Kaplan responded, "Well, this is a really bad time because we're waiting for the king of Sweden at the moment." The king interrupted him. "I am the king of Sweden." The king turned out to be perfectly tech savvy: he had a deep understanding of what they were trying to do, more so than most of their prospective customers—which, of course, was at the heart of the challenge that they faced.

There was, however, one distinct upside for Kaplan. He was invited to an evening reception for the king held at the Bohemian Club in San Francisco. He arrived and fell into conversation with a beautiful Swedish woman. They spoke for almost an hour and Kaplan thought that maybe she was the queen. As it turned out, she was a stewardess who worked for the Swedish airline that flew the royal entourage to the United States. The joke cut both ways, because she thought he was Steve Jobs. There was a happy ending. They would date for the next eight years.

Teknowledge wasn't so lucky. The company had a good dose of "The Smartest Guys in the Room" syndrome. With a who's who of some of the best engineers in AI, they had captured the magic of the new field and for what might otherwise pass for exorbitant consulting fees they would impart their alchemy. However, artificial intelligence systems at the time were little more than accretions of if-then-else statements packaged in overpriced workstations and presented with what were then unusually large computer displays with alluring graphical interfaces. In truth, they were more smoke and mirrors than canned expertise.

It was Kaplan himself who would become something of a Trojan horse within the company. In 1981 the IBM PC had legitimized personal computers and dramatically reduced their cost while expanding the popular reach of computing. Doug Engelbart and Alan Kay's intelligence augmentation—IA— meme was showing up everywhere. Computing could be used to extend or replace people, and the falling cost made it possible for software designers to take either path. Computing was now sneaking out from behind the carefully maintained glass wall of the corporate data center and showing up in the corporate office supplies budget.

Kaplan was quick to grasp the implications of the changes. Larry Tesler, a former SAIL researcher who would work for Steve Jobs in designing the Lisa and the Macintosh and help engineer the Newton for John Sculley, had the same early epiphany. He had tried to warn his coworkers at Xerox PARC that cheap PCs were going to change the world, but at the time—1975—no one was listening. Six years later, many people still didn't comprehend the impact of the falling cost of the microprocessor. Teknowledge's expert system software was then designed and deployed on an overpriced workstation, which cost about $17,000, and a complete installation might run between $50,000 and $100,000. But Kaplan realized that PCs were already powerful enough to run the high-priced Teknowledge software handily. Of course, the business implication was that without their flashy workstation trappings, they would be seen for what they really were—software packages that should sell for PC software prices.

Nobody at Teknowledge wanted to hear this particular heresy. So Kaplan did what he had done a few years earlier when he had briefly helped found Symantec in his spare time at Stanford. It was Christmas, and everyone else was on vacation, so he holed up in his cottage in the hills behind Stanford and went to work rewriting the Teknowledge software to run on a PC. Kaplan used a copy of Turbo Pascal, a lightning-fast programming language that made his version of the expert system interpreter run faster than the original workstation product. He finished the program over the holidays and came in and demoed the Wine Advisor, the Teknowledge demonstration program, on his "toy" personal computer. It just killed the official software running on the Xerox Star workstation.

All hell broke loose. Not only did it break the Teknowledge business model because software for personal computers was comparatively dirt cheap, but it violated their very sense of their place in the universe! Everyone hated him. Nonetheless, Kaplan managed to persuade Lee Hecht to commit to putting out a product based on the PC technology. But it was crazy. It meant selling a product for $80 rather than $80,000. Kaplan had become the apostate and he knew he was heading for the door. Ann Winblad, who was then working as a Wall Street technology analyst and would later become a well-known Silicon Valley venture capitalist, came by and Kaplan pitched her on the changing state of the computing world.

"I know someone you need to meet," she told him.

That someone turned out to be Mitch Kapor, the founder and chief executive of Lotus Development Corporation, the publisher of the 1-2-3 spreadsheet program. Kapor came by and Kaplan pitched him on his AI-for-the masses vision. The Lotus founder was enthusiastic about the idea: "I've got money, why don't you propose a product you want to build for me," he said.

Kaplan's first idea was to invent an inexpensive version of the Teknowledge expert system to be called ABC, as a play on 1-2-3. The idea attracted little enthusiasm. Not long afterward, however, he was flying on Kapor's private jet. The Lotus founder sat down with paper notes and a bulky Compaq computer the size of a sewing machine and began typing. That gave Kaplan a new idea. He proposed a free-form note-taking program that would act as a calendar and a repository for all the odds and ends of daily life. Kapor loved the idea and with Ed Belove, another Lotus software designer, the three men outlined a set of ideas for the program.

Kaplan again retreated back to his cottage, this time for a year and a half, just writing the code for the program with Belove while Kapor helped with the overall design. Lotus Agenda was the first of a new breed of packaged software, known as a Personal Information Manager, which was in some ways a harbinger of the World Wide Web. Information could be stored in free form and would be automatically organized into categories. It came to be described as a "spreadsheet for words" and it was a classic example of a new generation of software tools that empowered their users in the Engelbart tradition.

Introduced in 1988 to glowing reviews from industry analysts like Esther Dyson, it would go on to gather a cult following. The American AI Winter was just arriving and most of the new wave of AI companies would soon wilt, but Kaplan had been early to see the writing on the wall. Like Breiner, he went quickly from being an AI ninja to a convert to Engelbart's world of augmentation. PCs were the most powerful intellectual tool in history. It was becoming clear that it was equally possible to design humans into and out of systems being created with computers. Just as AI stumbled commercially, personal computing and thus intelligence augmentation shot ahead. In the late 1970s and early 1980s the personal computer industry exploded on the American scene. Overnight the idea that computing could be both a "fantasy amplifier" at home and a productivity tool at the office replaced the view of computing as an impersonal bureaucratic tool of governments and corporations. By 1982 personal computing had become such a cultural phenomenon that *Time* magazine put the PC on its cover as "Man of the Year."

It was the designers themselves who made the choice of IA over AI. Kaplan would go on to found Go Corp. and design the first pen-based computers that would anticipate the iPhone and iPad by more than a decade. Like Sheldon Breiner, who was also driven away from artificial intelligence by the 1980s AI Winter, he would become part of the movement toward human-centered design in a coming post-PC era.

The quest to build a working artificial intelligence was marked from the outset by false hopes and bitter technical and philosophical quarrels. In 1958, two years after the Dartmouth Summer Research Project on Artificial Intelligence, the *New York Times* published a brief UPI wire story buried on page 25 of the paper. The headline read NEW NAVY DEVICE LEARNS BY DOING: PSYCHOLOGIST SHOWS EMBRYO OF COMPUTER DESIGNED TO READ AND GROW WISER.[43]

The article was an account of a demonstration given by Cornell psychologist Frank Rosenblatt, describing the "embryo" of an electronic computer that the navy expected would one day "walk, talk, see, write, reproduce itself and be conscious of its existence." The device, at this point, was actually a simulation running on the Weather Bureau's IBM 704 computer that was able to tell right from left after some fifty attempts, according to the report. Within a year, the navy apparently was planning to build a "thinking machine" based on these circuits, for a cost of $100,000.

Dr. Rosenblatt told the reporters that this would be the first device to think "as the human brain," and that it would make mistakes initially but would grow wiser with experience. He suggested that one application for the new mechanical brain might be as a proxy for space exploration in lieu of humans. The article concluded that the first perceptron, an electronic or software effort to model biological neurons, would have about a thousand electronic "association cells" receiving electrical impulses from four hundred photocells—"eye-like" scanning devices. In contrast, it noted, the human brain was composed of ten billion responsive cells and a hundred million connections with the eyes.

The earliest work on artificial neural networks dates back to the 1940s, and in 1949 that research had caught the eye of Marvin Minsky, then a young Harvard mathematics student, who would go on to build early electronic learning networks, one as an undergraduate and a second one, named the Stochastic Neural Analog Reinforcement Calculator, or SNARC, as a graduate student at Princeton. He would later write his doctoral thesis on neural networks. These mathematical constructs are networks of nodes or "neurons" that are interconnected by numerical values that serve as "weights" or "vectors." They can be trained by being exposed to a series of patterns such as images or sounds to later recognize similar patterns.

During the 1960s a number of competing paths toward building thinking machines emerged, and the dominant direction became the logic- and rule-based approach favored by John McCarthy. However, during the same period, groups around the country were experimenting with competing analog approaches based on the earlier neural network ideas. It's ironic that Minsky, one of the ten attendees at the Dartmouth conference, would in 1969 precipitate a legendary controversy by writing the book *Perceptrons* with Seymour Papert, an analysis of neural networks that is widely believed to have stalled neural net research for many years. There is general agreement that as a consequence of the critique set forth in their book, the two MIT artificial intelligence researchers significantly delayed the young research area.

In fact, it was just one of a series of heated intellectual battles within the AI community during the sixties. Minsky and Papert have since argued that the criticism was unfair and that their book was a more balanced analysis of neural networks than was conceded by its critics. This dispute was further complicated by the fact that one of the main figures in the field, Rosenblatt, would die two years later in a sailing accident, leaving a vacuum in research activity into neural nets.

Early neural network research included work done at Stanford University as well as the research led by Charlie Rosen at SRI, but the Stanford group refocused its attention on telecommunications and Rosen would shift his Shakey work toward the dominant AI framework. Interest in neural networks would not reemerge until 1978, with the work of Terry Sejnowski, a postdoctoral student in neurobiology at Harvard. Sejnowski had given up his early focus on physics and turned to neuroscience. After taking a summer course in Woods Hole, Massachusetts, he found himself captivated by the mystery of the brain. That year a British postdoctoral psychologist, Geoffrey Hinton, was studying at the University of California at San Diego under David Rumelhart. The older UC scientist had created the parallel-distributed processing group with Donald Norman, the founder of the cognitive psychology department at the school.

Hinton, who was the great-great-grandson of logician George Boole, had come to the United States as a "refugee" as a direct consequence of the original AI Winter in England. The Lighthill report had asserted that most AI research had significantly underdelivered on its promise, the exception being computational neuroscience. In a Lighthill-BBC televised "hearing," both sides made their arguments based on the then state-of-the-art performance of computers. Neither side seemed to have taken note of the Moore's law of acceleration of computing speeds.

As a graduate student Hinton felt personally victimized by Minsky and Papert's attacks on neural networks. When he would tell people that he was working on artificial neural networks as a graduate student in England, their response would be, "Don't you get it? Those things are no good." His advisor told him to forget his interests and read Terry Winograd's thesis. It was all going to be symbolic logic in the future. But Hinton was on a different path. He was beginning to form a perspective that he would later describe as "neuro-inspired" engineering. He did not go to the extreme of some in the new realm of biological computing. He thought that slavishly copying biology would be a mistake. Decades later the same issue remains hotly disputed. In 2014 the European Union funded Swiss researcher Henry Markram with more than $1 billion to model a human brain at the tiniest level of detail, and Hinton was certain that the project was doomed to failure.

In 1982 Hinton had organized a summer workshop focusing on parallel models of associated memory, where Terry Sejnowski applied to attend. Independently the young physicist had been thinking about how the brain might be modeled using some of the new schemes that were being developed. It was the first scientific meeting that Hinton had organized. He was aware that the invited crowd had met repeatedly in the past—people he thought of as "elderly professors in their forties" would come and give their same old talks. He drew up a flyer and sent it to his targeted computer science and psychology departments. It offered to pay expenses for those with new ideas. He was predictably disappointed when most of the responses came at the problem using traditional approaches within computer science and psychology. But one of the proposals stood out. It was from a young scientist who claimed to have figured out the "machine code of the brain."

At roughly the same time Hinton was attending a conference with David Marr, the well-known MIT vision researcher, and he asked him if the guy was crazy. Marr responded that he knew him and that he was very bright and he had no idea if he was crazy or not. What was clear was that Sejnowski was pursuing a set of new ideas about cognition.

At the meeting Hinton and Sejnowski met for the first time. UCSD was already alive with a set of new ideas attempting to create new models of how the brain worked. Known as Parallel Distributed Processing, or PDP, it was a break from the symbol processing approach that was then dominating artificial intelligence and the cognitive sciences. They quickly realized they had been thinking about the problem from a similar perspective. They could both see the power of a new approach based on webs of sensors or "neurons" that were interconnected by a lattice of values representing connection strengths. In this new direction, if you wanted the network to interpret an image, you described the image in terms of a web of weighted connections. It proved to be a vastly more effective approach than the original symbolic model for artificial intelligence.

Everything changed in 1982 when Sejnowski's former physics advisor at Princeton, John Hopfield, invented what would become known as the Hopfield Network. Hopfield's approach broke from earlier neural network models that had been created by the designers of the first perceptrons, by allowing the individual neurons to update their values independently. The fresh approach to the idea of neural networks inspired both Hinton and Sejnowski to join in an intense collaboration.

The two young scientists had both taken their first teaching positions by that time, Hinton at Carnegie Mellon and

Sejnowski at Johns Hopkins, but they had become friends and were close enough that they could make the four-hour drive back and forth on weekends. They realized they had found a way to transform the original neural network model into a more powerful learning algorithm. They knew that humans learn by seeing examples and generalizing, and so mimicking that process became their focus. In creating a new kind of multilayered network, which they called a Boltzmann Machine, an homage to the Austrian physicist Ludwig Boltzmann. In their new model they conceived a more powerful approach to machine learning and made the most significant advance in design since the original single-layer learning algorithm designed by Rosenblatt.

Sejnowski had missed the entire political debate over the perceptron. As a physics graduate student he had been outside the world of artificial intelligence in the late 1960s when Minsky and Papert had made their attacks. Yet he had read the original *Perceptron* book and he had loved it for its beautiful geometric insights. He had basically ignored their argument that the perceptron would not be generalizable to the world of multilayer systems. Now he was able to prove them wrong.

Hinton and Sejnowski had developed an alternative model, but they needed to prove its power in contrast to the rule-based systems popular at the time. During the summer, with help of a graduate student, Sejnowski settled on a language problem to demonstrate the power of the new technique, training his neural net to pronounce English text as an alternative to a rule-based approach. At the time he had no experience in linguistics and so he went to the school library and checked out a textbook that was a large compendium of pronunciation rules. The book documented the incredibly complex set of rules and exceptions required to speak the English language correctly.

Halfway through their work on a neural network able to learn to pronounce English correctly, Hinton came to Baltimore for a visit. He was skeptical.

"This probably won't work," he said. "English is an incredibly complicated language and your simple network won't be able to absorb it."

So they decided to begin with a subset of the language. They went to the library again and found a children's book with a very small set of words. They brought up the network and set it to work absorbing the language in the children's book. It was spooky that within an hour it began to work. At first the sounds it generated were gibberish, like the sounds an infant might make, but as it was trained it improved continuously. Initially it got a couple of words correct and then it continued until it was able to perfect itself. It learned from both the general rules and the special cases.

They went back to the library and got another linguistics text containing a transcription of a story told by a fifth grader about what it was like in school and a trip to his grandmother's house on one side of the page. On the other side were the actual sounds for each word transcribed by a phonologist. It was a perfect teacher for their artificial neurons and so they ran that information through the neural network. It was a relatively small corpus, but the network began to speak just like the fifth grader. The researchers were amazed and their appetite was whetted.

Next they got a copy of a twenty-thousand-word dictionary and decided to see how far they could push their prototype neural network. This time they let the program run for a week on what was a powerful computer for its day, a Digital Equipment Corp. VAX minicomputer. It learned and learned and learned and ultimately it was able to pronounce new words it had never seen before. It was doing an amazingly good job.

They called the program Nettalk. It was built out of three hundred simulated circuits they called neurons. They were arranged in three layers—an input layer to capture the words, an output layer to generate the speech sounds, and a "hidden layer" to connect the two. The neurons were interconnected to one another by eighteen thousand "synapses"—links that had numeric values that could be represented as weights. If these simple networks could "learn" to hear, see, speak, and generally mimic the range of things that humans do, they were obviously a powerful new direction for both artificial intelligence and augmentation.

After the success of Nettalk, Sejnowski's and Hinton's careers diverged. Sejnowski moved to California and joined the Salk Institute, where his research focused on theoretical problems in neuroscience. In exploring the brain he became a deep believer in the power of diversity as a basic principle of biology—a fundamental divergence from the way modern digital computing evolved. Hinton joined the computer science department at the University of Toronto and over the next two decades he would develop the original Boltzmann Machine approach. From the initial supervised model, he found ways to add unsupervised (automatic) learning. The Internet became a godsend, providing vast troves of data in the form of crowd-sourced images, videos, and snippets of speech, both labeled and unlabeled. The advances would eventually underpin a dramatic new tool for companies like Google, Microsoft, and Apple that were anxious to deploy Internet services based on vision, speech, and pattern recognition.

This complete reversal of the perceptron's fate also lay in part in a clever public relations campaign, years in the making. Before Sejnowski and Hinton's first encounter in San Diego, a cerebral young French student, Yann LeCun, had stumbled across Seymour Papert's dismissive discussion of the perceptron, and it sparked his interest. After reading the account, LeCun headed to the library to learn everything he could about machines that were capable of learning. The son of an aerospace engineer, he had grown up tinkering with aviation hardware and was steeped in electronics before going to college. He would have studied astrophysics, but he enjoyed hacking too much. He read the entire literature on the perceptron going back to the fifties and concluded that there was no one working on the subject in the early 1980s. It was the heyday of expert systems and no one was writing about neural networks.

In Europe his journey began as a lonely crusade. As an undergraduate he would study electrical engineering, and he began his Ph.D. work with someone who had no idea about the topic he was focusing on. Then shortly after he began his graduate studies he stumbled across an obscure article on Boltzmann Machines by Hinton and Sejnowski. "I've got to talk to these guys!" he thought to himself. "They are the only people who seem to understand."

Serendipitously, it turned out that they were able to meet in the winter of 1985 in the French Alps at a scientific conference on the convergence of ideas in physics and neuroscience. Hopfield Networks, which served as an early model for human memory, had sparked a new academic community of interest. Although Sejnowski attended the meeting he actually missed LeCun's talk. It was the first time the young French scientist had presented in English, and he had been terrified, mostly because there was a Bell Laboratories physicist at the conference who often arrogantly shot down each talk with criticisms. The people that LeCun was sitting next to told him that was the Bell Labs style—either the ideas were subpar, or the laboratory's scientists had already thought of them. To his shock, when he gave his talk in broken English, the Bell Labs scientist stood up and endorsed it. A year later, Bell Labs offered LeCun a job.

Later in the meeting, LeCun cornered Sejnowski and the two scientists compared notes. The conversation would lead to the creation of a small fraternity of researchers who would go on to formulate a new model for artificial intelligence. LeCun finished his thesis work on an approach to training neural networks known as "back propagation." His addition made it possible to automatically "tune" the networks to recognize patterns more accurately.

After leaving school LeCun looked around France to find organizations that were pursuing similar approaches to AI. Finding only a small ministry of science laboratory and a professor who was working in a related field, LeCun obtained funding and laboratory space. His new professor told him, "I've no idea what you're doing, but you seem like a smart guy so I'll sign the papers." But he didn't stay long. First he went off to Geoff Hinton's neural network group at the University of Toronto, and when the Bell Labs offer arrived he moved to New Jersey, continuing to refine his approach known as convolutional neural nets, initially focusing on the problem of recognizing handwritten characters for automated mail-sorting applications. French-born Canadian Yoshua Bengio, a bright MIT-trained computer scientist, joined him at Bell Labs and worked on the character recognition software, and later on machine vision technology that would be used by the NCR Corporation to automatically read a sizable proportion of all the bank checks circulating in the world.

Yet despite their success, for years the neural network devotees were largely ignored by the mainstream of academic computer science. Thinking of themselves as the "three musketeers," Hinton, LeCun, and Bengio set out to change that. Beginning in 2004 they embarked on a "conspiracy"—in LeCun's words—to boost the popularity of the networks, complete with a rebranding campaign offering more alluring concepts of the technology such as "deep learning" and "deep belief nets." LeCun had by this time moved to New York University, partly for closer ties with neuroscientists and with researchers applying machine-learning algorithms to the problem of vision.

Hinton approached a Canadian foundation, the Canadian Institute for Advanced Research, for support to organize a research effort in the field and to hold several workshops each year. Known as the Neural Computation and Adaptive Percep-

Terry Sejnowski, Yann LeCun, and Geoffrey Hinton (*from left to right*), three scientists who helped revive artificial intelligence by developing biologically inspired neural network algorithms. (*Photo courtesy of Yann LeCun*)

tion project, it permitted him to handpick the most suitable researchers in the world across a range of fields stretching from neuroscience to electrical engineering. It helped crystallize a community of people interested in the neural network research.

This time they had something else going for them—the pace of computing power had accelerated, making it possible to build neural networks of vast scale, processing data sets orders of magnitude larger than before. It had taken almost a decade, but by then the progress, power, and value of the neural network techniques was indisputable. In addition to raw computing power, the other missing ingredient had been large data sets to use to train the networks. That would change rapidly with the emergence of the global Internet, making possible a new style of centralized computing power—cloud computing— as well as the possibility of connecting that capacity to billions of mobile sensing and computing systems in the form of smartphones. Now the neural networks could be easily trained on millions of digital images or speech samples readily available via the network.

As the success of their techniques became more apparent, Hinton began to receive invitations from different computer companies all looking for ways to increase the accuracy of a wide variety of consumer-oriented artificial intelligence services—speech recognition, machine vision and object recognition, face detection, translation and conversational systems. It seemed like the list was endless. As a consultant Hinton had introduced the deep learning neural net approach early on at Microsoft, and he was vindicated in 2012, when Microsoft's head of research Richard Rashid gave a lecture in a vast auditorium in Tianjin, China. As the research executive spoke in English he paused after each sentence, which was then immediately translated by software into spoken Chinese in a simulation of his own voice. At the end of the talk, there was silence and then stunned applause from the audience.

The demonstration hadn't been perfect, but by adding deep learning algorithm techniques the company had adopted from Hinton's research, it had been able to reduce recognition errors by more than 30 percent. The following year a trickle of interest in neural networks turned into a torrent. The easy availability of Internet data sets and low-cost crowd-sourced labor provided both computing and human resources for training purposes.

Microsoft wasn't alone. A variety of new neural net and other machine-learning techniques have led to a dramatic revival of interest in AI in Silicon Valley and elsewhere. Combining the new approach to AI with the Internet has meant that it is now possible to create a new service based on computer vision or speech recognition and then use the Internet and tens of millions of smartphone users to immediately reach a global audience.

In 2010 Sebastian Thrun had come to Google to start the Google X Laboratory, which was initially framed inside the company as Google's version of Xerox's Palo Alto Research

Center. It had a broad portfolio of research projects, stretching from Thrun's work in autonomous cars to efforts to scale up neural networks, loosely identified as "brain" projects, evoking a new wave of AI.

The Human Brain Project was initially led by Andrew Ng, who had been a colleague with Thrun at the resurrected Stanford Artificial Intelligence Laboratory. Ng was an expert in machine learning and adept in some of the deep learning neural network techniques that Hinton and LeCun had pioneered. In 2011, he began spending time at Google building a machine vision system and the following year it had matured to the point where Google researchers presented a paper on how the network performed in an unsupervised learning experiment using YouTube videos. Training itself on ten million digital images found on YouTube, it performed far better than any previous effort by roughly doubling accuracy in recognizing objects from a challenging list of twenty thousand distinct items. It also taught itself to recognize cats, which is not surprising since there is an overabundance of cat images on YouTube. The Google brain assembled a dreamlike digital image of a cat by employing a hierarchy of memory locations to successively cull out general features after being exposed to millions of images. The scientists described the mechanism as a cybernetic cousin to what takes place in the brain's visual cortex. The experiment was made possible by Google's immense computing resources that allowed the researchers to turn loose a cluster of sixteen thousand processors on the problem—which of course is still a tiny fraction of the brain's billions of neurons, a huge portion of which are devoted to vision.

Whether or not Google is on the trail of a genuine artificial "brain" has become increasingly controversial. There is certainly no question that the deep learning techniques are paying off in a wealth of increasingly powerful AI achievements in vision and speech. And there remains in Silicon Valley a growing group of engineers and scientists who believe they are once again closing in on "Strong AI"—the creation of a self-aware machine with human or greater intelligence.

Ray Kurzweil, the artificial intelligence researcher and barnstorming advocate for technologically induced immortality, joined Google in 2013 to take over the brain work from Ng, shortly after publishing *How to Create a Mind,* a book that purported to offer a recipe for creating a working AI. Kurzweil, of course, has all along been one of the most eloquent backers of the idea of a singularity. Like Moravec, he posits a great acceleration of computing power that would lead to the emergence of autonomous superhuman machine intelligence, in Kurzweil's case pegging the date to sometime around 2023. The idea became codified in Silicon Valley in the form of the Singularity University and the Singularity Institute, organizations that focused on dealing with the consequences of that exponential acceleration.

Joining Kurzweil are a diverse group of scientists and engineers who believe that once they have discovered the mechanism underlying the biological human neuron, it will be simply a matter of scaling it up to create an AI. Jeff Hawkins, a successful Silicon Valley engineer who had founded Palm Computing with Donna Dubinsky, coauthored *On Intelligence* in 2004, which argued that the path to human-level intelligence lay in emulating and scaling up neocortex-like circuits capable of pattern recognition. In 2005, Hawkins formed Numenta, one of a growing list of AI companies pursuing pattern recognition technologies. Hawkins's theory has parallels with the claims that Kurzweil makes in *How to Create a Mind,* his 2012 effort to lay out a recipe for intelligence. Similar paths have been pursued by Dileep George, a Stanford-educated artificial intelligence researcher who originally worked with Hawkins at Numenta and then left to form his own company, Vicarious, with the goal of developing "the next generation of AI algorithms," and Henry Markram, the Swiss researcher who has enticed the European Union into supporting his effort to build a detailed replica of the human brain with one billion euros in funding.

In 2013 a technology talent gold rush that was already under way reached startling levels. Hinton left for Google because the resources available in Mountain View dwarfed what he had access to at the University of Toronto. There is now vastly more computing power available than when Sejnowski and Hinton first developed the Boltzmann Machine approach to neural networks, and there is vastly more data to train the networks on. The challenge now is managing a neural network that might have one billion parameters. To a conventional statistician that's a nightmare, but it has spawned a sprawling "big data" industry that does not shy away from monitoring and collecting virtually every aspect of human behavior, interaction, and thought.

After his arrival at Google, Hinton promptly published a significant breakthrough in making more powerful and efficient learning networks by discovering how to keep the parameters from effectively stepping on each other's toes. Rather than have an entire network process the whole image simultaneously, in the new model a subset is chosen, a portion of the image is processed, and the weights of the connections are updated. Then another random set is picked and the image is processed again. It offers a way to use randomness to reinforce the influence of each subset. The insight might be biologically inspired, but it's not a slavish copy. By Sejnowski's account, Hinton is an example of an artificial intelligence researcher who pays attention to the biology but is not constrained by it.

In 2012 Hinton's networks, trained on a huge farm of computers at Google, did remarkably well at recognizing individual objects, but they weren't capable of "scene understanding." For example, the networks could not recognize the sentence: "There is a cat sitting on the mat and there is a person dangling a toy at the cat." The holy grail of computer vision requires what AI researchers call "semantic understanding"—the ability to interpret the scene in terms of human language. In the 1970s the challenge of scene understanding was strongly influenced by Noam Chomsky's ideas about generative grammar as a context for objects and a structure for understanding their relation within a scene. But for decades the research went nowhere.

However, late in 2014, the neural network community began to make transformative progress in this domain as well. Around the country research groups reported progress in combining the learning properties of two different types of neural networks, one to recognize patterns in human language and the other to recognize patterns in digital images. Strikingly, they produced programs that could generate English-language sentences that described images at a high level of abstraction.[44] The advance will help in applications that improve the results generated by Internet image search applications. The new approach also holds out the potential for creating a class of programs that can interact with humans with a more sophisticated level of understanding.

Deep learning nets have made significant advances, but for Hinton, the journey is only now beginning. He said recently that he sees himself as an explorer who has landed on a new continent and it's all very interesting, but he has only progressed a hundred yards inland and it's still looking very interesting—except for the mosquitoes. In the end, however, it's a new continent and the researchers still have no idea what is really possible.

In late 2013, LeCun followed Hinton's move from academia to industry. He agreed to set up and lead Facebook's AI research laboratory in New York City. The move underscored the renewed corporate enthusiasm for artificial intelligence. The AI Winter was only the dimmest of memories. It was now clearly AI Spring.

Facebook's move to join the AI gold rush was an odd affair. It began with a visit by Mark Zuckerberg, Facebook cofounder and chief executive, to an out-of-the-way technical conference called Neural Information Processing Systems, or NIPS, held in a Lake Tahoe hotel at the end of 2013. The meeting had long been a bone-dry academic event, but Zuckerberg's appearance to answer questions was a clear bellwether. Not only were the researchers unused to appearances by high-visibility corporate tycoons, but Zuckerberg was accompanied by uniformed guards, lending the event a surreal quality. The celebrity CEO filled the room he was in and several other workshops were postponed as a video feed was piped into an overflow room. "The tone changed rapidly: accomplished professors became little more than lowly researchers shuffling into the Deep Learning workshop to see a Very Important Person speak,"[45] blogged Alex Rubinsteyn, a machine-learning researcher who was an attendee at the NIPS meeting.

In the aftermath of the event there was an alarmed back-and-forth inside the tiny community of researchers about the impact of commercialization of AI on the culture of the academic research community. It was, however, too late to turn back. The field has moved on from the intellectual quarrels in the 1950s and 1960s over the feasibility of AI and the question of the correct path. Today, a series of probabilistic mathematical techniques have reinvented the field and transformed it from an academic curiosity into a force that is altering many aspects of the modern world.

It has also created an increasingly clear choice for designers. It is now possible to design humans into or out of the computerized systems that are being created to grow our food, transport us, manufacture our goods and services, and entertain us. It has become a philosophical and ethical choice, rather than simply a technical one. Indeed, the explosion of computing power and its accessibility everywhere via wireless networks has reframed with new urgency the question addressed so differently by McCarthy and Engelbart at the dawn of the computing age.

In the future will important decisions be made by humans or by the deep learning–style algorithms? Today, the computing world is demarcated between those who focus on creating intelligent machines and those who focus on how human capabilities can be extended by the same machines. Will it surprise anyone that the differing futures emerging from those opposing stances must be very different worlds?

5 | WALKING AWAY

As a young navy technician in the 1950s, Robert Taylor had a lot of experience flying, even without a pilot's license. He had become a favorite copilot of the real pilots, who needed both lots of flight hours and time to study for exams. So they would take Taylor along in their training jets, and after they took off he would fly the plane—gently—while the real pilots studied in the backseat. He even practiced instrument landing approaches, in which the plane is guided to a landing by radio communications, while the pilot wears a hood blocking the view of the outside terrain.

As a young NASA program administrator in the early 1960s, Taylor was confident when he received an invitation to take part in a test flight at a Cornell University aerospace laboratory. On arrival they put him in an uncomfortable anti-g suit and plunked him down in the front seat of a Lockheed T-33 jet trainer while the real pilot sat behind him. They took off and the pilot flew up to the jet's maximum altitude, almost fifty thousand feet, then Taylor was offered the controls to fly around a bit. After a while the pilot said, "Let's try something a little more interesting. Why don't you put the plane in a dive?" So Taylor pushed the joystick forward until he thought he was descending steeply enough, then he began to ease the stick backward. Suddenly he froze in panic. As he pulled back, the plane entered a steeper dive. It felt like going over the top of a roller-coaster ride. He pulled the stick back farther but the plane was still descending almost vertically.

Finally he said to the pilot behind him, "Okay, you've got it, you better take over!" The pilot laughed, leveled the plane out, and said, "Let's try this again." They tried again and this time when he pushed the stick forward the plane went unexpectedly upward. As he pushed a bit harder, the plane tilted up farther. This time he panicked, about to stall the plane, and again the pilot leveled the plane out.

Taylor should have guessed. He had had such an odd piloting experience because he was unwittingly flying a laboratory plane that the air force researchers were using to experiment with flight control systems. The air force invited Taylor to Cornell because as a NASA program manager he had granted, unsolicited, $100,000 to a flight research group at Wright-Patterson Air Force Base.

Taylor, first at NASA and then at DARPA, would pave the way for systems used both to augment humans and to replace them. NASA was three years old when, in 1961, President Kennedy had announced the goal of getting an American to the moon—and back—safely during that decade. Taylor found himself at an agency with a unique charter, to fundamentally shape how humans and machines interact, not just in flight, but ultimately in all computer-based systems from the desktop PC to today's mobile robots.

The term "cyborg," for "cybernetic organism," had been coined originally in 1960 by medical researchers thinking about intentionally enhancing humans to prepare them for the exploration of space.[1] They foresaw a new kind of creature—half human, half mechanism—capable of surviving in harsh environments.

In contrast, Taylor's organization was funding the design of electronic systems that closely collaborated with humans while retaining a bright line distinguishing what was human and what was machine.

In the early 1960s NASA was a brand-new government bureaucracy deeply divided by the question of the role of humans in spaceflight. For the first time it was possible to conceive of entirely automated flight in space. The deeply unsettling idea was an obvious future direction—one in which machines would steer and humans would be passengers, then already the default approach pursued by the Soviet space program. In contrast, the U.S. program underscored the deep division that was highlighted by a series of incidents where American astronauts had intervened, thus proving the survival value of what in NASA parlance came to be called "human in the loop." On *Gemini VI,* for example, Wally Schirra was hailed as a hero after he held off pushing the abort button during a launch sequence, even though he was violating a NASA mission rule.[2]

The human-in-the-loop debates became a series of intensely fought battles inside NASA during the 1950s and 1960s. When Taylor arrived at the agency in 1961 he found an engineering culture in love with a body of mathematics known as control theory, Norbert Wiener's cybernetic legacy. These NASA engineers were designing the nation's aeronautic as well as astronautic flight systems. These were systems of such complexity that the engineers found them abstractly, some might say inherently, beautiful. Taylor could see early on that the aerospace designers were wedded to the aesthetics of control as much as the fact that the systems needed to be increasingly automated because humans weren't fast or reliable enough to control them.

He had stumbled into an almost intractable challenge, and

hence a deeply divided technical culture. NASA was split on the question of the role of humans in spaceflight. Taylor saw that the dispute pervaded even the highest echelons of the agency, and that it was easy to predict which side of the debate each particular manager would take. Former jet pilots would be in favor of keeping a human in the system, while experts in control theory would choose full automation.

As a program manager in 1961, Taylor was responsible for several areas of research funding, one of them called "manned flight control systems." Another colleague in the same funding office was responsible for "automatic control systems." The two got along well enough, but they were locked in a bitter budgetary zero-sum game. Taylor began to understand the arguments his colleagues made in support of automated control, though he was responsible for mastering arguments for manned control. His best card in the debate was that he had the astronauts on his side and they had tremendous clout. NASA's corps of astronauts had mostly been test pilots. They were the pride of the space agency and proved Taylor's invaluable allies. Taylor had funded the design and construction of simulator technology used extensively in astronaut training—systems for practicing a series of spacecraft maneuvers, like docking—since the early days of the Mercury program, and had spent hours talking with astronauts about the strengths and weaknesses of the different virtual training environments. He found that the astronauts were keenly aware of the debate over the proper role of humans in the space programs. They had a huge stake in whether they would have a role in future space systems or be little more than another batch of dogs and smart monkeys coming along for the ride.

The political battle over the human in the loop was waged over two divergent narratives: that of the heroic astronauts landing on the surface of the moon and that of the specter of a catastrophic accident culminating in the deaths of the astronauts—and potentially, as a consequence, the death of the agency. The issue, however, was at least temporarily settled when during the first human moon landing Neil Armstrong heroically took command after a computer malfunction and piloted the *Apollo 11* spacecraft safely to the lunar surface. The moon landing and other similar feats of courage, such as Wally Schirra's decision not to abort the earlier Gemini flight, have firmly established a view of human-machine interaction that elevates human decision-making beyond the fallible machines of our mythology. Indeed, the macho view of astronauts as modern-day Lewises and Clarks was from the beginning deeply woven into the NASA ethos, as well as being a striking contrast with the early Soviet decision to train women cosmonauts.[3] The American view of human-controlled systems was long partially governed by perceived distinctions between U.S. and Soviet approaches to aeronautics as well as astronautics. The Vostok spacecraft were more automated, and so Soviet astronauts were basically passengers rather than pilots. Yet the original American commitment to human-controlled spaceflight was made when aeronautical technology was in its infancy. In the ensuing half century, computers and automated systems have become vastly more reliable.

For Taylor, the NASA human-in-the-loop wars were a formative experience that governed his judgment at both NASA and DARPA, where he projected and sponsored technological breakthroughs in computing, robotics, and artificial intelligence. While at NASA, Taylor fell into the orbit of J. C. R. Licklider, whose interests in psychology and information technology led him to anticipate the full potential of interactive computing. In his seminal 1960 paper "Man-Computer Symbiosis," Licklider foresaw an era when computerized systems would entirely displace humans. However, he also predicted an interim period that might span from fifteen to five hundred years in which humans and computers would cooperate. He believed that that period would be the most "intellectually and most creative and exciting [time] in the history of mankind."

Taylor moved to ARPA in 1965 as Licklider's protégé. He set about funding the ARPAnet, the first nationwide research-oriented computer network. In 1968 the two men coauthored a follow-up to Licklider's symbiosis paper titled "The Computer as a Communication Device." In it, Licklider and Taylor were possibly the first to delineate the coming impact of computer networks on society.

Today, even after decades of research in human-machine and human-computer interaction in the airplane cockpit, the argument remains unsettled—and has emerged again with the rise of autonomous navigation in trains and automobiles. While Google leads in research in driverless cars, the legacy automobile industry has started to deploy intelligent systems that can offer autonomous driving in some well-defined cases, such as during stop-and-go traffic jams, but then return the car to human control in situations recognized as too complex or risky to autopilot. It may take seconds for a human sitting in the driver's seat, possibly distracted by an email or worse, to return to "situational awareness" and safely resume control of the car. Indeed the Google researchers may have already come up against the limits to autonomous driving. There is currently a growing consensus that the "handoff" problem—returning manual control of an autonomous car to a human in the event of an emergency—may not actually be a solvable one. If that proves true, the development of the safer cars of the future will tend toward augmentation technology rather than automation technology. Completely autonomous driving might ultimately be limited to special cases like low-speed urban services and freeway driving.

Nevertheless, the NASA disputes were a harbinger of the emerging world of autonomous machines. During the first fifty years of interactive computing, beginning in the mid-sixties, computers largely augmented humans instead of replacing them. The technologies that became the hallmark of Silicon Valley—personal computing and the Internet—largely amplified human intellect, although it was undeniably the case that an "augmented" human could do the work of several (former) coworkers. Today, in contrast, system designers have a choice. As AI technologies including vision, speech, and reasoning have begun to mature, it is increasingly possible to design humans either in or out of "the loop."

Funded first by J. C. R. Licklider and then, beginning in 1965, by Bob Taylor, John McCarthy and Doug Engelbart worked in laboratories just miles apart from each other at the outset of the modern computing era. They might as well have been in different universes. Both were funded by ARPA, but they had little if any contact. McCarthy was a brilliant, if somewhat cranky, mathematician and Engelbart was an Oregon farm boy and a dreamer.

The outcome of their competing pioneering research was unexpected. When McCarthy came to Stanford to create the Stanford Artificial Intelligence Laboratory in the mid-1960s, his work was at the very heart of computer science, focusing on big concepts like artificial intelligence and proof of software program correctness using formal logic. Engelbart, on the other hand, set out to build a "framework" for augmenting the human intellect. It was initially a more nebulous concept viewed as far outside the mainstream of academic computer science, and yet for the first three decades of the interactive computing era Engelbart's ideas had more worldly impact. Within a decade both the first modern personal computers and then later information-sharing technologies like the World Wide Web—both of which can be traced in part to Engelbart's research—emerged.

Since then Engelbart's adherents have transformed the world. They have extended human capabilities everywhere in modern life. Today, shrunk into smartphones, personal computers will soon be carried by all but the allergic or iconoclastic adult and teenager. Smartphones are almost by definition assembled into a vast distributed computing fabric woven together by the wireless Internet. They are also relied on as artificial memories. Today many people are literally unable to hold a conversation or find their way around town without querying them.

While Engelbart's original research led directly to the PC and the Internet, McCarthy's lab was most closely associated with two other technologies—robotics and artificial intelligence. There had been no single dramatic breakthrough. Rather, the falling cost of computing (both in processing and storage), the gradual shift from the symbolic logic-based approach of the first generation of AI research to more pragmatic statistics and machine-learning algorithms of the second generation of AI, and the declining price of sensors now offer engineers and programmers the canvas to create computerized systems that see, speak, listen, and move around in the world.

The balance has shifted. Computing technologies are emerging that can be used to replace and even outpace humans. At the same time, in the ensuing half century there has been little movement toward unification in the two fields, IA and AI, the offshoots of Engelbart's and McCarthy's original work. Rather, as computing and robotics systems have grown from laboratory curiosities into the fabric that weaves together modern life, the opposing viewpoints of those in each community have for the most part continued to speak past each other.

The human-computer interaction community keeps debating metaphors ranging from windows and mice to autonomous agents, but has largely operated within the philosophical framework originally set down by Engelbart—that computers should be used to augment humans. In contrast, the artificial intelligence community has for the most part pursued performance and economic goals elaborated in equations and algorithms, largely unconcerned with defining or in any way preserving a role for individual humans. In some cases the impact is easily visible, such as manufacturing robots that directly replace human labor. In other cases it is more difficult to discern the direct effect on employment caused by deployment of new technologies. Winston Churchill said: "We shape our buildings, and afterwards our buildings shape us." Today our systems have become immense computational edifices that define the way we interact with our society, from how our physical buildings function to the very structure of our organizations, whether they are governments, corporations, or churches.

As the technologies marshaled by the AI and IA communities continue to reshape the world, alternative visions of the future play out: In one world humans coexist and prosper with the machines they've created—robots care for the elderly, cars drive themselves, and repetitive labor and drudgery vanish, creating a new Athens where people do science, make art, and enjoy life. It will be wonderful if the Information Age unfolds in that fashion, but how can it be a foregone conclusion? It is equally possible to make the case that these powerful and productive technologies, rather than freeing humanity, will instead facilitate a further concentration of wealth, fomenting vast new waves of technological unemployment, casting an inescapable surveillance net around the globe, while unleashing a new generation of autonomous superweapons.

When Ed Feigenbaum finished speaking the room was silent. No polite applause, no chorus of boos. Just a hush. Then the conference attendees filed out of the room and left the artificial intelligence pioneer alone at the podium.

Shortly after Barack Obama was elected president in 2008, it seemed possible that the Bush administration plan for space exploration, which focused on placing a manned base on the moon, might be replaced with an even more audacious program that would involve missions to asteroids and possibly even manned flights to Mars with human landings on the Martian moons Phobos and Deimos.[4] Shorter-term goals included the possibility of sending astronauts to Lagrangian points one million miles from Earth where the Earth's and Sun's gravitational pull cancel each other and create convenient long-term parking for ambitious devices like a next-generation Hubble Space Telescope.

Human exploration of the solar system was the pet project of G. Scott Hubbard, a head of NASA's Ames Research Center in Mountain View, California, who was heavily backed by the Planetary Society, a nonprofit that advocates for space exploration and science. As a result, NASA organized a conference to discuss the possible resurrection of human exploration of the solar system. A star-studded cast of space luminaries, including astronaut Buzz Aldrin, the second human to set foot on the moon, and celebrity astrophysicist Neil deGrasse Tyson, showed up for the day. One of the panels focused on the role of robots, which were envisioned by the conference organizers as providing intelligent systems that would assist humans on long flights to other worlds.

Feigenbaum had been a student of one of the founders of the field of AI, Herbert Simon, and he had led the development of the first expert systems as a young professor at Stanford. A believer in the potential of artificial intelligence and robotics, he had been irritated by a past run-in with a Mars geologist who had insisted that sending a human to Mars would provide more scientific information in just a few minutes than a complete robot mission might return. Feigenbaum also had a deep familiarity with the design of space systems. Moreover, having once served as chief scientist of the air force, he was a veteran of the human-in-the-loop debates stretching back to the space program.

He showed up to speak at the panel with a chip on his shoulder. Speaking from a simple set of slides, he sketched out

an alternative to the manned flight to Mars vision. He rarely
used capital letters in his slides, but he did this time:

ALMOST EVERYTHING THAT HAS BEEN LEARNED
ABOUT THE SOLAR SYSTEM AND SPACE BEYOND HAS
BEEN LEARNED BY PEOPLE <u>ON EARTH</u> ASSISTED BY
THEIR NHA (NON-HUMAN AGENTS) IN SPACE OR IN
ORBIT[5]

The whole notion of sending humans to another planet when
robots could perform just as well—and maybe even better—for
a fraction of the cost and with no risk of human life seemed
like a fool's errand to Feigenbaum. His point was that AI systems and robots in the broader sense of the term were becoming so capable so quickly that the old human-in-the-loop idea
had lost its mystique as well as its value. All the coefficients on
the nonhuman side of the equation had changed. He wanted to
persuade the audience to start thinking in terms of agents, to
shift gears and think about humans exploring the solar system
with augmented senses. It was not a message that the audience
wanted to hear. As the room emptied, a scientist who worked
at NASA's Goddard Space Flight Center came to the table and
quietly said that she was glad that Feigenbaum had said what
he did. In her job, she whispered, she could not say that.

Feigenbaum's encounter underscores the reality that there
isn't a single "right" answer in the dichotomy between AI and
IA. Sending humans into space is a passionate ideal for some.
For others like Feigenbaum, however, the vast resources the
goal entails are wasted. Intelligent machines are perfectly
suited for the hostile environment beyond Earth, and in designing them we can perfect technologies that can be used to good
effect on Earth. His quarrel is also indicative that there won't
be any easy synthesis of the two camps.

While the separate fields of artificial intelligence and
human-computer interaction have largely remained isolated

domains, there are people who have lived in both worlds and researchers who have famously crossed from one camp to the other. Microsoft cognitive psychologist Jonathan Grudin first noted that the two fields have risen and fallen in popularity, largely in opposition to each other. When the field of artificial intelligence was more prominent, human-computer interaction generally took a backseat, and vice versa.

Grudin thinks of himself as an optimist. He has written that he believes it is possible that in the future there will be a grand convergence of the fields. Yet the relationship between the two fields remains contentious and the human-computer interaction perspective as pioneered by Engelbart and championed by people like Grudin and his mentor Donald Norman is perhaps the most significant counterweight to artificial intelligence–oriented technologies that have the twin potential for either liberating or enslaving humanity.

While Grudin has oscillated back and forth between the AI and IA worlds throughout his career, Terry Winograd became the first high-profile deserter from the world of AI. He chose to walk away from the field after having created one of the defining software programs of the early artificial intelligence era and has devoted the rest of his career to human-centered computing, or IA. He crossed over.

Winograd's interest in computing was sparked while he was a junior studying math at Colorado College, when a professor of medicine asked his department for help doing radiation therapy calculations.[6] The computer available at the medical center was a piano-sized Control Data minicomputer, the CDC 160A, one of Seymour Cray's first designs. One person at a time used it, feeding in programs written in Fortran by way of a telex-like punched paper tape. On one of Winograd's first days using the machine, it was rather hot so there was a fan sitting behind the desk that housed the computer terminal. He managed to feed his paper tape into the computer and then, by mistake, right into the fan.[7]

Terry Winograd was a brilliant young graduate student at MIT who developed an early program capable of processing natural language. Years later he rejected artificial intelligence research in favor of human-centered software design. (*Photo courtesy of Terry Winograd*)

In addition to his fascination with computing, Winograd had become intrigued by some of the early papers about artificial intelligence. As a math whiz with an interest in linguistics, the obvious place for graduate studies was MIT. When he arrived, at the height of the Vietnam War, Winograd discovered there was a deep gulf between the rival fiefdoms of Marvin Minsky and Noam Chomsky, leaders in the respective fields of artificial intelligence and linguistics. The schism was so deep that when Winograd would bump into Chomsky's students at parties and mention that he was in the AI Lab, they would turn and walk away.

Winograd tried to bridge the gap by taking a course from Chomsky, but he received a C on a paper in which he argued for the AI perspective. Despite the conflict, it was a heady time for AI research. The Vietnam War had opened the Pentagon's research coffers and ARPA was essentially writing blank checks to researchers at the major research laboratories. As at Stanford, at MIT there was a clear sense of what "serious" research in computer science was about. Doug Engelbart came around on a tour and showed a film demonstration of his NLS system. The researchers at the MIT AI Lab belittled his accomplishments. After all, they were building systems that would soon have capabilities matching those of humans, and Engelbart was showing off a computer editing system that seemed to do little more than sort grocery lists.

At the time Winograd was very much within the mainstream of computing, and as the zeitgeist pointed toward artificial intelligence, he followed. Most believed that it wouldn't be long before machines would see, hear, speak, move, and otherwise perform humanlike tasks. Winograd was soon encouraged to pursue linguistic research by Minsky, who was eager to prove that his students could do as well or better at "language" than Chomsky's. That challenge was fine with Winograd, who was interested in studying how language worked by using computing as a simulation tool.

As a teenager growing up in Colorado, Winograd, like many of his generation, had discovered *Mad* magazine. The irreverent—and frequently immature—satire journal would play a small role in naming SHRDLU, a program he wrote as a graduate student at MIT in the late 1960s that "understood" natural language and responded to commands. It has remained one of the most influential artificial intelligence programs.

Winograd had set out to build a system that could respond to typed commands in natural language and perform useful tasks in response. By this time there had already been a wave of initial experiments in building conversational programs. Eliza, written by MIT computer scientist Joseph Weizenbaum in 1964 and 1965, was named after Eliza Doolittle, who learned proper English in Shaw's *Pygmalion* and the musical *My Fair Lady*. Eliza had been a groundbreaking experiment in the study of human interaction with machines: it was one of the first programs to provide users the opportunity to have a humanlike conversation with a computer. In order to skirt the need for real-world knowledge, Eliza parroted a Rogerian therapist and frequently reframed users' statements as questions. The conversation was mostly one-sided because Eliza was programmed simply to respond to certain key words and phrases. This approach led to wild non sequiturs and bizarre detours. For example, Eliza would respond to a user's statement about their mother with: "You say your mother?" Weizenbaum later said that he was stunned to discover Eliza users became deeply engrossed in conversations with the program, and even revealed intimate personal details. It was a remarkable insight not into the nature of machines but rather into human nature. Humans, it turns out, have a propensity to find humanity in almost everything they interact with, ranging from inanimate objects to software programs that offer the illusion of human intelligence.

Was it possible that in the cyber-future, humans, increasingly isolated from each other, would remain in contact with some surrogate computer intelligence? What kind of world did that foretell? Perhaps it was the one described in the movie *Her*, released in 2013, in which a shy guy connects with a female AI. Today, however, it is still unclear whether the emergence of cyberspace is a huge step forward for humanity as described by cyber-utopians such as Grateful Dead lyricist John Perry Barlow in his 1996 *Wired* manifesto, "A Declaration of the Independence of Cyberspace," or the much bleaker world described by Sherry Turkle in her book *Alone Together: Why We Expect More from Technology and Less from Each Other*. For Barlow, cyberspace would become a utopian world free from crime and degradation of "meatspace." In contrast, Turkle describes a world in which computer networks increasingly drive a wedge between humans, leaving them lonely and isolated. For Weizenbaum, computing systems risked fundamentally diminishing the human experience. In very much the same vein that Marxist philosopher Herbert Marcuse attacked advanced industrial society, he was concerned that the approaching Information Age might bring about a "One-Dimensional Man."

In the wake of the creation of Eliza, a group of MIT scientists, including information theory pioneer Claude Shannon, met in Concord, Massachusetts, to discuss the social implications of the phenomenon.[8] The seductive quality of the interactions with Eliza concerned Weizenbaum, who believed that an obsessive reliance on technology was indicative of a moral failing in society, an observation rooted in his experiences as a child growing up in Nazi Germany. In 1976, he sketched out a humanist critique of computer technology in his book *Computer Power and Human Reason: From Judgment to Calculation*. The book did not argue against the possibility of artificial intelligence but rather was a passionate indictment of computerized systems that substituted automated decision-making for the human mind. In the book, he argued that computing served as a conservative force in society by propping up bureaucracies as well as by reductively redefining the world as a narrow and more sterile place by restricting the potential of human relationships.

Weizenbaum's criticism largely fell on deaf ears in the United States. Years later his ideas would receive a more positive reception in Europe, where he moved at the end of his life. At the time, however, in the United States, where the new computing technologies were taking root, there was more optimism about artificial intelligence.

In the late 1960s as a graduate student, Winograd was immersed in the hothouse world of the MIT AI Lab, the birthplace of the computing hacker culture, which would lead both to personal computing and the "information wants to be free" ideology that would later become the foundation of the open-source computing movement of the 1990s. Many at the lab staked their careers on the faith that cooperative and autonomous intelligent machines would soon be a reality. Eliza, and then several years later Winograd's SHRDLU, were the direct predecessors of the more sophisticated computerized personal assistants that would follow in the coming decades. There had been earlier efforts at MIT to build microworlds or "block worlds," which were restricted, simulated environments in which AI researchers would create programs capable of reasoning about their surroundings and planning. Some of those environments had used real robot arms and blocks. When Winograd began working on his project, another student was already building a system that could book airline reservations, but that was less interesting to Winograd. Instead, he set out to build a constrained world to explore and rationalize and chose to create his system in the form of a virtual computer world.

He built a computer simulation as a world populated by colored blocks that could be explored and manipulated by an artificial intelligence programming language named Micro-Planner, based on the work of Carl Hewitt, another MIT graduate student. Given the relatively primitive state of computing at the time, much was left to the imagination. There was no fancy graphical animation of the blocks world. The user simply sat at a Teletype terminal and entered questions at the keyboard and the computer responded, in natural language. Winograd chose the name SHRDLU because no better one came to mind. He later said the sequence SHRDLU was probably hidden in his subconscious. The sequence "etaoin shrdlu," representing the most frequently used letters in English, are on the left-hand side of the Linotype keyboard, as "1qaz 2wsx" appear on the typewriter—and as a consequence the string of characters showed up occasionally in *Mad* magazine, as an inside joke on the accidental appearance of the character string in print. The Linotype, which cast an entire line of type from molten metal, had no backspace. Because it was faster to cast an erroneous line than correct it, typesetters would "run down" the rest of the line with easy-to-type nonsense, later removing the entire line after it had cooled down, or if they forgot, hope a proof-reader caught it.[9]

He wasn't concerned at the time about any ethical implications involved in building a natural language processing system that could "understand" and respond in a virtual world. In SHRDLU "understanding" meant that the program analyzed the structure of the typed questions and attempted to answer them and respond to the commands. It was an early effort at disambiguation, a thorny problem for natural language processing even today. For example, in the sentence "he put the glass on the table and it broke," does "it" refer to the glass or the table? Without more context, neither a human nor an AI program could decide.

Winograd's system used its general knowledge of the microworld to answer and respond to various questions. He illustrated the conversation as follows in his dissertation:

Person: Pick up a big red block.
Computer: OK.
Person: Grasp the pyramid.
Computer: I don't understand which pyramid you mean.
Person (changing their mind): Find a block which is taller than the one you are holding and put it into the box.
Computer: By "it," I assume you mean the block which is taller than the one I am holding.
Computer: OK.
Person: What does the box contain?
Computer: The blue pyramid and the blue block.
Person: What is the pyramid supported by?
Computer: The box.
Person: How many blocks are not in the box?
Computer: Four of them.
Person: Is at least one of them narrower than the one which I told you to pick up?
Computer: Yes, the red cube.[10]

In 1970, when it was completed, SHRDLU proved to be one of the most successful demonstrations of AI's potential—perhaps too successful. Years later, Winograd's tour de force would be blamed for helping generate the optimistic view that it would be possible to "scale up" similar programs to deal with real-world complexity. For example, during the 1980s and 1990s the AI research community widely accepted that it would be possible to build a machine with the reasoning power at least of a kindergartener—by simply accumulating a vast number of common-sense rules.

The attack on the AI optimists, however, had begun even before Winograd built SHRDLU. Although Weizenbaum's critique was about the morality of building intelligent machines, the more heated debate was over whether such machines were even possible. Seymour Papert, Winograd's thesis advisor, had become engaged in a bitter debate with Hubert Dreyfus, a philosopher and Heidegger acolyte, who, just one decade after McCarthy had coined the term, would ridicule the field in a scathing paper entitled "Alchemy and Artificial Intelligence," published in 1965 by the RAND Corporation.[11] (Years later, in the 2014 movie remake of *RoboCop,* the fictional U.S. senator who sponsors legislation banning police robots is named Hubert Dreyfus in homage.)

Dreyfus ran afoul of AI researchers in the early sixties when they showed up in his Heidegger course and belittled philosophers for failing to understand human intelligence after studying it for centuries.[12] It was a slight he would not forget. For the next four decades, Dreyfus would become the most pessimistic critic of the possibility of work-as-promised artificial intelligence, summing up his argument in an attack on two Stanford AI researchers: "Feigenbaum and Feldman claim that tangible progress is indeed being made, and they define progress very carefully as 'displacement toward the ultimate goal.' According to this definition, the first man to climb a tree could claim tangible progress toward flight to the moon."[13] Three years later, Papert fired back in "The Artificial Intelligence of Hubert L. Dreyfus, A Budget of Fallacies": "The perturbing observation is not that Dreyfus imports metaphysics into engineering but that *his discussion is irresponsible,*" he wrote. "His facts are almost always wrong; his insight into programming is so poor that he classifies as impossible programs a beginner could write; and his logical insensitivity allows him to take *his* inability to imagine how a *particular* algorithm can be carried out, as reason to believe no algorithm can achieve the desired purpose."[14]

Winograd would eventually break completely with Papert, but this would not happen for many years. He came to Stanford as a professor in 1973, when his wife, a physician, accepted an offer as a medical resident in the Bay Area. It was just two years after Intel had introduced the first commercial 4004 microprocessor chip, and trade journalist Don Hoefler settled on "Silicon Valley U.S.A." as shorthand for the region in his newsletter *Microelectronics News.* Winograd continued to work for several years on the problem of machine understanding of natural language very much in the original tradition of SHRDLU. Initially he spent almost half his time at Xerox Palo Alto Research Center working with Danny Bobrow, another AI researcher interested in natural language understanding. Xerox had opened a beautiful new building in March 1975 in a location next to Stanford, as it gave the "document company" easy access to the best computer scientists. Later Winograd would tell friends, "You know all the famous personal computing technology that was invented at PARC? Well, that's not what I worked on."

Instead he spent his time trying to elaborate and expand on the research he pursued at MIT, research that would bear fruit almost four decades later. During the 1970s, however, it seemed to present an impossible challenge, and many started to wonder how, or even if, science could come to understand how humans process language. After spending a half decade

on language-related computing Winograd found himself growing more and more skeptical that real progress in AI would
be possible. In addition to making little headway, he rejected
artificial intelligence in part because of the influence of a new
friendship with a Chilean political refugee named Fernando
Flores, and in part because of his recent engagement with a
group of Berkeley philosophers, led by Dreyfus, intent on strip-
ping away the hype around the new AI industry now emerging.
Flores, a bona fide technocrat who had been finance minister
during the Allende government, barely escaped his office in the
palace when it was bombed during the coup. He spent three
years in prison before arriving in the United States, his release
coming in response to political pressure by Amnesty International. Stanford had appointed Flores as a visiting scholar
in computer science, but he left Palo Alto instead to pursue a
Ph.D. at Berkeley under the guidance of a quartet of anti-AI
philosophers: Hubert and Stuart Dreyfus, John Searle, and
Ann Markussen.

Winograd thought Flores was one of the most impressive
intellectuals he had ever met. "We started talking in a casual
way, then he handed me a book on philosophy of science and
said, 'You should read this.' I read it, and we started talking about it, and we decided to write a paper about it, that
turned into a monograph, and that turned into a book. It was
a gradual process of finding him interesting, and finding the
stuff we were talking about intellectually stimulating," Winograd recalled.[15] The conversations with Flores put the young
computer scientist "in touch" with the ways in which he was
unhappy with what he thought of as the "ideology" of AI.
Flores aligned himself with the charismatic Werner Erhard,
whose cultlike organization EST (Erhard Seminars Training)
had a large following in the Bay Area during the 1970s. (At
Stanford Research Institute, Engelbart sent the entire staff of
his lab through EST training and joined the board of the organization.)

Although the computing world was tiny at the time, the tensions between McCarthy and Minsky's AI design approach and Engelbart's IA approach were palpable around Stanford. PARC was inventing the personal computer; the Stanford AI Lab was doing research on everything from robot arms to mobile robots to chess-playing AI systems. At the recently renamed SRI (which changed its name from Stanford Research Institute due to student antiwar protests) researchers were working on projects that ranged from Engelbart's NLS system to Shakey the robot, as well as early speech recognition research and "smart" weapons. Winograd would visit Berkeley for informal lunchtime discussions with Searle and Dreyfus, the Berkeley philosophers, their grad students, and Fernando Flores. While Hubert Dreyfus objected to the early optimistic predictions by AI researchers, it was John Searle who raised the stakes and asked one of the defining philosophical questions of the twentieth century: Is it possible to build an intelligent machine?

Searle, a dramatic lecturer with a flair for showmanship, was never one to avoid an argument. Before teaching philosophy he had been a political activist. While at the University of Wisconsin in the 1950s he had been a member of Students Against Joseph McCarthy, and in 1964 he would become the first tenured Berkeley faculty to join the Free Speech Movement. As a young philosopher Searle had been drawn to the interdisciplinary field of cognitive science. At the time, the core assumption of the field was that the biological mind was analogous to the software that animated machines. If this was the case, then understanding the processes of human thought would merely be a matter of teasing out the program inside the intertwined billions of neurons making up the human brain.

The Sloan Foundation had sent Searle to Yale to discuss the subject of artificial intelligence. While on the plane to the meeting he began reading a book about artificial intelligence written by Roger Schank and Robert Abelson, the leading Yale AI researchers during the second half of the 1970s. *Scripts,*

Plans, Goals, and Understanding[16] made the assertion that artificial intelligence programs could "understand" stories that had been designed by their developers. For example, developers could present the computer with a simple story, such as a description of a man going into a restaurant, ordering a hamburger, and then storming out without paying for it. In response to a query, the program was able to infer that the man had not eaten the hamburger. "That can't be right," Searle thought to himself, "because you could give me a story in Chinese with a whole lot of rules for shuffling the Chinese symbols, and I don't understand a word of Chinese but all the same I could give the right answer."[17] He decided that it just didn't follow that the computer had the ability to understand anything just because it could interpret a set of rules.

While flying to his lecture, he came up with what has been called the "Chinese Room" argument against sentient machines. Searle's critique was that there could be no simulated "brains in a box." His argument was different from the original Dreyfus critique, which asserted that obtaining human-level performance from AI software was impossible. Searle simply argued that a computing machine is little more than a very fast symbol shuffler that uses a set of syntactical rules. What it lacks is what the biological mind has—the ability to interpret semantics. The biological origin of semantics, the formal study of meaning, remains a great mystery. Searle's argument was infuriating to the AI community in part because he implied that their argument implicitly linked them with a theological argument that the mind is outside the physical, biological world. His argument was that mental processes are entirely caused by biological processes in the brain and they are realized there, and if you want to make a machine that can think, you must duplicate, rather than simulate, those processes. At the time Searle thought that they had probably already considered his objection and the discussion wouldn't last a week, let alone decades. But it has. Searle's original article generated thirty published refutations. Three decades later, the debate is anything but settled. To date, there are several hundred published attacks on his idea. And Searle is still alive and busy defending his position.

It is also notable that the lunchtime discussions about the possibility of intelligent and conceivably self-aware machines took place against a backdrop of the Reagan military buildup. The Vietnam War had ended, but there were still active pockets of political dissent around the country. The philosophers would meet at the Y across the street from the Berkeley campus. Winograd and Danny Bobrow from Xerox PARC had become regular visitors at these lunches, and Winograd found that they challenged his intellectual biases about the philosophical underpinnings of AI.

He would eventually give up the AI "faith." Winograd concluded that there was nothing mystical about human intelligence. In principle, if you could discover the way the brain worked, you could build a functional artificially intelligent machine, but you couldn't build that same machine with symbolic logic and computing, which was the dominant approach in the 1970s and 1980s. Winograd's interest in artificial intelligence had been twofold: AI served both as a model for understanding language and the human brain and as a system that could perform useful tasks. At that point, however, he took an "Engelbartian" turn. Philosophically and politically, human-centered computing was a better fit with his view of the world. Winograd had gotten intellectually involved with Flores, which led to a book, *Understanding Computers and Cognition: A New Foundation for Design,* a critique of artificial intelligence. *Understanding Computers,* though, was philosophy, not science, and Winograd still had to figure out what to do with his career. Eventually, he set down his effort to build smarter machines and focused instead on the question of how to use computers to make people smarter. Winograd crossed the chasm. From designing systems that were intended to supplant humans he turned his focus to working on technologies that enhanced the way people interact with computers.

Though Winograd would argue years later that politics had not directly played a role in his turn away from artificial intelligence, the political climate of the time certainly influenced many other scientists' decisions to abandon the artificial intelligence camp. During a crucial period from 1975 to 1985, artificial intelligence research was overwhelmingly funded by the Defense Department. Some of the nation's most notable computer scientists—including Winograd—had started to worry about the increasing involvement of the military in computing technology R & D. For a generation who had grown up watching the movie *Dr. Strangelove,* the Reagan administration Star Wars antimissile program seemed like dangerous brinkmanship. It was at least a part of Winograd's moral background and was clearly part of the intellectual backdrop during the time when he decided to leave the field he had helped to create. Winograd was a self-described "child of the '60s,"[18] and during the crucial years when he turned away from AI, he simultaneously played a key role in building a national organization of computer scientists, led by researchers at Xerox PARC and Stanford, who had become alarmed at the Star Wars weapons buildup. The group shared a deep fear that the U.S. military command would push the country into a nuclear confrontation with the Soviet Union. As a graduate student Winograd had been active against the war in Vietnam while he was in Boston as part of a group called "Computer People for Peace." In 1981 he became active again as a leader in helping create a national organization of computer scientists who opposed nuclear weapons.

In response to the highly technical Strategic Defense Initiative, the disaffected computer scientists believed they could use the weight of their expertise to create a more effective anti–nuclear weapons group. They evolved from being "people" and became "professionals." In 1981, they founded a new organization called Computer Professionals for Social Responsibility. Winograd ran the first planning meeting, held in a large classroom at Stanford. Those who attended recalled that unlike many political meetings from the antiwar era that were marked by acrimony and debate, the evening was characterized by an unusual sense of unity and common purpose. Winograd proved an effective political organizer.

In a 1984 essay on the question of whether computer scientists should accept military funding, Winograd pointed out that he had avoided applying for military funding in the past, but by keeping his decision private, he had ducked what he would come to view as a broader responsibility. He had, of course, received his training in a military-funded laboratory at MIT. Helping establish Computer Professionals for Social Responsibility was the first of a set of events that would eventually lead Winograd to "desert" the AI community and turn his attention from building intelligent machines to augmenting humans.

Indirectly it was a move that would have a vast impact on the world. Winograd was recognized enough in the artificial intelligence community that, if he had decided to pursue a more typical academic career, he could have built an academic empire based on his research interests. Personally, however, he had no interest in building a large research lab or even supporting postdoctoral researchers. He was passionate about one-to-one interaction with his students.

One of these was Larry Page, a brash young man with a wide range of ideas for possible dissertation topics. Under Winograd's guidance Page settled on the idea of downloading the entire Web and improving the way information was organized and discovered. He set about doing this by mining human knowledge, which was embodied in existing Web hyperlinks. In 1998, Winograd and Page joined with Sergey Brin, another Stanford graduate student and a close friend of

Page's, and Brin's faculty advisor, Rajeev Motwani, an expert in data mining, to coauthor a journal article titled "What Can You Do with a Web in Your Pocket?"[19] In the paper, they described the prototype version of the Google search engine.

Page had been thinking about other more conventional AI research ideas, like self-driving cars. Instead, with Winograd's encouragement, he would find an ingenious way of mining human behavior and intelligence by exploiting the links created by millions of Web users. He used this information to significantly improve the quality of the results returned by a search engine. This work would be responsible for the most significant "augmentation" tool in human history. In September of that year, Page and Brin left Stanford and founded Google, Inc. with the modest goal of "organizing the world's knowledge and making it universally useful."

By the end of the 1990s Winograd believed that the artificial intelligence and human-computer interaction research communities represented fundamentally different philosophies about how computers and humans should interact. The easy solution, he argued, would be to agree that both camps were equally "right" and to stipulate that there will obviously be problems in the world that could be solved by either approach. This answer, however, would obscure the fact that inherent in these differing approaches are design consequences that play out in the nature of the systems. Adherents of the different philosophies, of course, construct these systems. Winograd had come to believe that the way computerized systems are designed has consequences both in how we understand humans and how technologies are designed for their benefit.

The AI approach, which Winograd describes as "rationalistic," views people as machines. Humans are modeled with internal mechanisms very much like digital computers. "The key assumptions of the rationalistic approach are that the essential aspects of thought can be captured in a formal symbolic representation," he wrote. "Armed with this logic, we can create intelligent programs and we can design systems that optimize human interaction."[20] In opposition to the rational AI approach was the augmentation method that Winograd describes as "design." That approach is more common in the human-computer interaction community, in which developers focus not on modeling a single human intelligence, but rather on using the relationship between the human and the environment as the starting point for their investigations, be it with humans or an ensemble of machines. Described as "human-centered" design, this school of thought eschews formal planning in favor of an iterative approach to design, encapsulated well in the words of industrial designer and IDEO founder David Kelley: "Enlightened trial and error outperforms the planning of flawless intellect."[21] Pioneered by psychologists and computer scientists like Donald Norman at the University of California at San Diego and Ben Shneiderman at the University of Maryland, human-centered design would become an increasingly popular approach that veered away from the rationalist AI model that was popularized in the 1980s.

In the wake of the defeats of the AI Winter in the 1980s, in the 1990s, the artificial intelligence community also changed dramatically. It largely abandoned its original formal, rationalist, top-down straitjacket that had been described as GOFAI, or "Good Old-Fashioned Artificial Intelligence," in favor of statistical and "bottom-up," or "constructivist," approaches, such as those pursued by roboticists led by Rod Brooks. Nevertheless, the two communities have remained distant, preoccupied with their contradictory challenges of either replacing or augmenting human skills.

In breaking with the AI community, Winograd became a member of a group of scientists and engineers who took a step back and rethought the relationship between humans and the smart tools they were building. In doing so, he also reframed the concept of "machine" intelligence. By posing the question of whether humans were actually "thinking machines" in the same manner of the computing machines that the AI researchers were trying to create, he argued that the very question makes us engage—wittingly or not—in an act of projection that tells us more about our concept of human intelligence than it does about the machines we are trying to understand. Winograd came to believe that intelligence is an artifact of our social nature, and that we flatten our humaneness by simplifying and distorting what it is to be human as simulated by a machine.

While artificial intelligence researchers rarely spoke to the human-centered design researchers, the two groups would occasionally organize confrontational sessions at technical conferences. In the 1990s, Ben Shneiderman was a University of Maryland computer scientist who had become a passionate advocate of the idea of human-centered design through what became known as "direct manipulation." During the 1980s, with the advent of Apple's Macintosh and Microsoft's Windows software systems, direct manipulation had become the dominant style in computer user interfaces. For example, rather than entering commands on a keyboard, users could change the shape of an image displayed on a computer screen by grabbing its edges or corners with a mouse and dragging them.

Shneiderman was at the top of his game and, during the 1990s, he was a regular consultant at companies like Apple, where he dispensed advice on how to efficiently design computer interfaces. Shneiderman, who considered himself to be an opponent of AI, counted among his influences Marshall McLuhan. During college, after attending a McLuhan lecture at the Ninety-Second Street Y in New York City, he had felt emboldened to pursue his own various interests, which crossed the boundaries between science and the humanities. He went home and printed a business card describing his job title as "General Eclectic" and subtitled it "Progress is not our most important product."[22]

He would come to take pride in the fact that Terry Winograd had moved from the AI camp to the HCI world. Shneiderman sharply disagreed with Winograd's thesis when he read it in the 1970s and had written a critical chapter about SHRDLU in his 1980 book *Software Psychology.* Some years later, when Winograd and Flores published *Understanding Computers and Cognition,* which made the point that computers were unable to "understand" human language, he called Winograd up and told him, "You were my enemy, but I see you've changed." Winograd laughed and told Shneiderman that *Software Psychology* was required reading in his classes. The two men became good friends.

In his lectures and writing, Shneiderman didn't mince words in his attacks on the AI world. He argued not only that the AI technologies would fail, but also that they were poorly designed and ethically compromised because they were not designed to help humans. With great enthusiasm, he argued that autonomous systems raised profound moral issues related to who was responsible for the actions of the systems, issues that weren't being addressed by computer researchers. This fervor wasn't new for Shneiderman, who had previously been involved in legendary shouting matches at technical meetings over the wisdom of designing animated human agents like Microsoft Clippy, the Office assistant, and Bob, the ill-received attempts Microsoft made to design more "friendly" user interfaces.

In the early 1990s anthropomorphic interfaces had become something of a fad in computer design circles. Inspired in part by Apple's widely viewed Knowledge Navigator video, computer interface designers were adding helpful and chatty animated cartoon figures to systems. Banks were experimenting with animated characters that would interact with customers from the displays of automated teller machines, and car manufacturers started to design cars with speech synthesis that would, for example, warn drivers when their door was ajar. The initial infatuation would come to an abrupt halt, however, with the embarrassing failure of Microsoft Bob. Although it had been designed with the aid of Stanford University user interface specialists, the program was widely derided as a goofy idea.

Did the problem with Microsoft Bob lie with the idea of a "social" interface itself, or instead with the way it was implemented? Microsoft's bumbling efforts were rooted in the work of Stanford researchers Clifford Nass and Byron Reeves, who had discovered that humans responded well to computer interfaces that offered the illusion of human interaction. The two researchers arrived at the Stanford Communications Department simultaneously in 1986. Reeves had been a professor of communications at the University of Wisconsin, and Nass had studied mathematics at Princeton and worked at IBM and Intel before turning his interests toward sociology.

As a social scientist Nass worked with Reeves to conduct a series of experiments that led to a theory of communications they described as "the Media Equation." In their book, *The Media Equation: How People Treat Computers, Television, and New Media Like Real People and Places,* they explored what they saw as the human desire to interact with technological devices—computers, televisions, and other electronic media—in the same "social" fashion with which they interacted with other humans. After writing *The Media Equation,* Reeves and Nass were hired as consultants for Microsoft in 1992 and encouraged the design of familiar social and natural interfaces. This extended the thinking underlying Apple's graphical interface for the Macintosh, which, like Windows, had been inspired by the original work done on the Alto at Xerox PARC. Both were designs that attempted to ease the task of using a computer by creating a graphical environment that was evocative of a desk and office environment in the physical world. However, Microsoft Bob, which attempted to extend the "desktop" metaphor by creating a graphical computer environment that evoked the family home, adopted a cartoonish and dumbed-down approach that the computer digerati found insulting to users, and the customer base overwhelmingly rejected it.

Decades later the success of Apple's Siri has vindicated Nass and Reeves's early research, suggesting that the failure of Microsoft Bob lay in how Microsoft built and applied the system rather than in the approach itself. Siri speeds people up in contexts where keyboard input might be difficult or unsafe, such as while walking or driving. Both Microsoft Bob and Clippy, on the other hand, slowed down user engagement with the program and came across as overly simplistic and condescending to users: "as if they were being asked to learn to ride a bicycle by starting with a tricycle," according to Tandy Trower, a veteran Microsoft executive.[23] That said, Trower pointed out that Microsoft may have fundamentally mistaken the insights offered by the Stanford social scientists: "Nass and Reeves' research suggests that user expectations of human-like behavior are raised as characters become more human," he wrote. "This Einstein character sneezed when you asked it to exit. While no users were ever sprayed upon by the character's departure, if you study Nass and Reeves, this is considered to be socially inappropriate and rude behavior. It doesn't matter that they are just silly little animations on the screen; most people still respond negatively to such behavior."[24]

Software agents had originally emerged during the first years of the artificial intelligence era when Oliver Selfridge and his student Marvin Minsky, both participants in the original Dartmouth AI conference, proposed an approach to machine perception called "Pandemonium," in which collaborative programs called "demons," described as "intelligent agents," would work in parallel on a computer vision problem. The original software agents were merely programs that ran inside a computer. Over two decades computer scientists, science-fiction authors, and filmmakers embellished the idea. As it evolved, it became a powerful vision of an interconnected, computerized world in which software programs cooperated in pursuit of a common goal. These programs would collect information, perform tasks, and interact with users as animated servants. But was there not a Faustian side to this? Shneiderman worried that leaving computers to complete human tasks would create more problems than it solved. This concern was at the core of his attack on the AI designers.

Before their first debate began, Shneiderman tried to defuse the tension by handing Pattie Maes, who had recently become a mother, a teddy bear. At two technical meetings in 1997 he squared off against Maes over AI and software agents. Maes was a computer scientist at the MIT Media Lab who, under the guidance of laboratory founder Nicholas Negroponte, had started developing software agents to perform useful tasks on behalf of a computer user. The idea of agents was just one of many future-of-computing ideas pursued at Negroponte's laboratory, which started out as ArcMac, the Architecture Machine Group, and groomed multiple generations of researchers who took the lab's "demo or die" ethos to heart. His original ArcMac research group and its follow-on, the MIT Media Laboratory, played a significant role in generating many of the ideas that would filter into computing products at both Apple and Microsoft.

In the 1960s and 1970s, Negroponte, who had trained as an architect, traced a path from the concept of a human-machine design partnership to a then far-out vision of "architecture without architects" in his books *The Architecture Machine* and *Soft Architecture Machines.*

In his 1995 book *Being Digital,* Negroponte, a close friend to AI researchers like Minsky and Papert, described his view of the future of human-computer interaction: "What we today call 'agent-based interfaces' will emerge as the dominant means by which computers and people talk with one another."[25] In 1995, Maes founded Agents, Inc., a music recommendation service, with a small group of Media Lab partners. Eventually the company would be sold to Microsoft, which used the privacy technologies her company had developed but did not commercialize its original software agent ideas.

At first the conference organizers had wanted Shneiderman and Maes to debate the possibility of artificial intelligence. Shneiderman declined and the topic was changed. The two researchers agreed to debate the contrasting virtues of software agents that acted on a user's behalf, on the one hand, and software technologies that directly empowered a computer user, on the other.

The high-profile debate took place in March of 1997 at the Association for Computing Machinery's Computers and Human Interaction (CHI) Conference in Atlanta. The event was given top billing along with other questions of pressing concern like, "Why Aren't We All Flying Personal Helicopters?" and "The Only Good Computer Is an Invisible Computer?" In front of an audience of the world's best computer interface designers, the two computer scientists spent an hour laying out the pros and cons of designs that directly augment humans and those that work more or less independently of them.

"I believe the language of 'intelligent autonomous agents' undermines human responsibility," Shneiderman said. "I can show you numerous articles in the popular press which suggest the computer is the active and responsible party. We need to clarify that either programmers or operators are the cause of computer failures."[26]

Maes responded pragmatically. Shneiderman's research was in the Engelbart tradition of building complex systems to give users immense power, and as a result they required significant training. "I believe that there are real limits to what we can do with visualization and direct manipulation because our computer environments are becoming more and more complex," she responded. "We cannot just add more and more sliders and buttons. Also, there are limitations because the users are not computer-trained. So, I believe that we will have to, to some extent, delegate certain tasks or certain parts of tasks to agents that can act on our behalf or that can at least make suggestions to us."[27]

Perhaps Maes's most effective retort was that it might be wrong to believe that humans always wanted to be in control and to be responsible. "I believe that users sometimes want to be couch-potatoes and wait for an agent to suggest a movie for them to look at, rather than using 4,000 sliders, or however many it is, to come up with a movie that they may want to see," she argued. Things politely concluded with no obvious winner, but it was clear to Jonathan Grudin, who was watching from the audience, that Pattie Maes had been brave to debate this at a CHI conference, on Shneiderman's home turf. The debate took place a decade and a half before Apple unveiled Siri, which successfully added an entirely artificial human element to human-computer interaction. Years later Shneiderman would acknowledge that there were some cases in which using speech and voice recognition might be appropriate. He did, however, remain a staunch critic of the basic idea of software agents, and pointed out that aircraft cockpit designers had for decades tried and failed to use speech recognition to control airplanes.

When Siri was introduced in 2010, the "Internet of Things" was approaching the peak in the hype cycle. This had originally been Xerox PARC's next big idea after personal computing. In the late 1980s PARC computer scientist Mark Weiser had predicted that as microprocessor cost, size, and power collapsed, it would be possible to discreetly integrate computer intelligence into everyday objects. He called this "UbiComp" or ubiquitous computing. Computing would disappear into the woodwork, he argued, just as electric motors, pulleys, and belts are now "invisible." Outside Weiser's office was a small sign: UBICOMP IS UPWARDLY COMPATIBLE WITH REALITY. (A popular definition of "ubiquitous" is "notable only for its absence.")

It would be Steve Jobs who once again most successfully took advantage of PARC's research results. In the 1980s he had borrowed the original desktop computing metaphor from PARC to design the Lisa and then the Macintosh computers. Then, a little more than a decade later, he would be the first to successfully translate Xerox's ubiquitous computing concept for a broad consumer audience. The iPod, first released in October of 2001, was a music player reconceptualized for the ubiquitous computing world, and the iPhone was a digital transformation of the telephone. Jobs also understood that while Clippy and Bob were tone deaf on the desktop, on a mobile phone, a simulated human assistant made complete sense.

Shneiderman, however, continued to believe that he had won the debate handily and that the issue of software agents had been put to bed.

I n a restored brick building in Boston, a humanoid figure turned its head. The robot was no more than an assemblage of plastic, motors, and wires, all topped by a movable flat LCD screen with a cartoon visage of eyes and eyebrows. Yet the distinctive motion elicited a small shock of recognition and empathy from an approaching human. Even in a stack of electronic boxes, sensors, and wires, the human mind has an uncanny ability to recognize the human form.

Meet Baxter, a robot designed to work alongside human workers that was unveiled with some fanfare in the fall of 2012. Baxter is relatively ponderous and not particularly dexterous. Instead of moving around on wheels or legs, it sits in one place on an inflexible fixed stand. Its hands are pincers capable of delicately picking up objects and putting them down. It is capable of little else. Despite its limitations, however, Baxter represents a new chapter in robotics. It is one of the first examples of Andy Rubin's credo that personal computers are sprouting legs and beginning to move around in the environment. Baxter is the progeny of Rodney Brooks, whose path to building helper robots traces directly from the founders of artificial intelligence.

McCarthy and Minsky went their separate ways in 1962, but the Stanford AI Laboratory where McCarthy settled attracted a ragtag crowd of hackers, a mirror image of the original MIT AI Lab remaining under Minsky's guidance. In 1969 the two labs were electronically linked via the ARPAnet, a precursor of the modern Internet, thus making it simple for researchers to share information. It was the height of the Vietnam War and artificial intelligence and robotics were heavily funded by the military, but the SAIL ethos was closer to the countercultural style of the San Francisco's Fillmore Auditorium than it was to the Pentagon on the Potomac.

Hans Moravec, an eccentric young graduate student, was camping in the attic of SAIL, while working on the Stanford Cart, an early four-wheeled mobile robot. A sauna had been installed in the basement, and psychodrama groups shared the lab space in the evenings. Available computer terminals displayed the message "Take me, I'm yours." "The Prancing Pony"—a fictional wayfarer's inn in Tolkien's *Lord of the Rings*—was a mainframe-connected vending machine selling food suitable for discerning hackers. Visitors were greeted in a small lobby decorated with an ungainly "You Are Here" mural echoing the famous Saul Steinberg *New Yorker* cover depicting a relativistic view of the most important place in the United States. The SAIL map was based on a simple view of the laboratory and the Stanford campus, but lots of people had added their own perspectives to the map, ranging from placing the visitor at the center of the human brain to placing the laboratory near an obscure star somewhere out on the arm of an average-sized spiral galaxy.

It provided a captivating welcome for Rodney Brooks, another new Stanford graduate student. A math prodigy from

Adelaide, Australia, raised by working-class parents, Brooks had grown up far from the can-do hacker culture in the United States. However, in 1969—along with millions of others around the world—he saw Kubrick's *2001: A Space Odyssey*. Like Jerry Kaplan, Brooks was not inspired to train to be an astronaut. He was instead seduced by HAL, the paranoid (or perhaps justifiably suspicious) AI.

Brooks puzzled about how he might create his own AI, and arriving at college, he had his first opportunity. On Sundays he had solo access to the school's mainframe for the entire day. There, he created his own AI-oriented programming language and designed an interactive interface on the mainframe display.[1] Brooks now went on to writing theorem proofs, thus unwittingly working in the formal, McCarthy-inspired artificial intelligence tradition. Building an artificial intelligence was what he wanted to do with his life.

Looking at a map of the United States, he concluded that Stanford was the closest university to Australia with an artificial intelligence graduate program and promptly applied. To his surprise, he was admitted. By the time of his arrival in the fall of 1977, the pulsating world of antiwar politics and the counterculture was beginning to wane in the Bay Area. Engelbart's group at SRI had been spun off, with his NLS system augmentation technology going to a corporate time-sharing company. Personal computing, however, was just beginning to turn heads—and souls—on the Midpeninsula. This was the heyday of the Homebrew Computer Club, which held its first meeting in March of 1975, the very same week the new Xerox PARC building opened. In his usual inclusive spirit McCarthy had invited the club to meet at his Stanford laboratory, but he remained skeptical about the idea of "personal computing." McCarthy had been instrumental in pioneering the use of mainframe computers as shared resources, and in his mental calculus it was wasteful to own an underpowered computer that would sit idle most of the time. Indeed, McCarthy's time-sharing ideas had developed from this desire to use computing systems more efficiently while conducting AI research. Perhaps in a display of wry humor, he placed a small note in the second Homebrew newsletter suggesting the formation of the "Bay Area Home Terminal Club," chartered to provide shared access on a Digital Equipment Corp. VAX mainframe computer. He thought that seventy-five dollars a month, not including terminal hardware and communications connectivity costs, might be a reasonable fee. He later described PARC's Alto/Dynabook design prototype—the template for all future personal computers—as "Xerox Heresies."

Alan Kay, who would become one of the main heretics, passed through SAIL briefly during his time teaching at Stanford. He was already carrying his "interim" Dynabook around and happily showing it off: a wooden facsimile preceding laptops by more than a decade. Kay hated his time in McCarthy's lab. He had a very different view of the role of computing, and his tenure at SAIL felt like working in the enemy's camp.

Alan Kay had first envisioned the idea of personal computing while he was a graduate student under Ivan Sutherland at the University of Utah. Kay had seen Engelbart speak when the SRI researcher toured the country giving demonstrations of NLS, the software environment that presaged the modern desktop PC windows-and-mouse environment. Kay was deeply influenced by Engelbart and NLS, and the latter's emphasis on boosting the productivity of small groups of collaborators—be they scientists, researchers, engineers, or hackers. He would take Engelbart's ideas a step further. Kay would reinvent the book for an interactive age. He wrote about the possibilities of "Personal Dynamic Media," inspiring the look and feel of the portable computers and tablets we use today. Kay believed personal computers would become a new universal medium, as ubiquitous as the printed page was in the 1960s and 1970s.

Like Engelbart's, Kay's views were radically different from those held by McCarthy's researchers at SAIL. The labs were not antithetical to each other, but there was a significant difference in emphasis. Kay, like Engelbart, put the human user at the center of his design. He wanted to build technologies to extend the intellectual reach of humans. He did, however, differ from Engelbart in his conception of cyberspace. Engelbart thought the intellectual relation between humans and information could be compared to driving a car; computer users would sail along an information highway. In contrast, Kay had internalized McLuhan's insight that "the medium is the message." Computing, he foresaw, would become a universal, overarching medium that would subsume speech, music, text, video, and communications.

Neither of those visions found traction at SAIL. Les Earnest, brought to SAIL by ARPA officials in 1965 to provide management skills that McCarthy lacked, has written that many of the computing technologies celebrated as coming out of SRI and PARC were simultaneously designed at SAIL. The difference was one of philosophy. SAIL's mission statement had originally been to build a working artificial intelligence in the span of a decade—perhaps a robot that could match wits with a human while physically exceeding their strength, speed, and dexterity. Generations of SAIL researchers would work toward systems supplanting rather than supplementing humans.

When Rod Brooks arrived at Stanford in the fall of 1977, McCarthy was already three years overdue on his ten-year goal for creating a working AI. It had also been two years since Hans Moravec fired his first broadside at McCarthy, arguing that exponentially growing computing power was the baseline ingredient to consider in artificial intelligence systems development. Brooks, whose Australian outsider's sensibility offered him a different perspective into the goings-on at Stanford, would become Moravec's night-shift assistant. Both had their quirks. Moravec was living at SAIL around the clock and counted on friends to bring him groceries. Brooks, too, quickly adopted the countercultural style of the era. He had shoulder-length hair and experimented with a hacker life-style: he worked a "28-hour day," which meant that he kept a 20-hour work-cycle, followed by 8 hours of sleep. The core thrust of Brooks's Ph.D. thesis, on symbolic reasoning about visual objects, followed in the footsteps of McCarthy. Beyond that, however, the Australian was able to pioneer the use of geometric reasoning in extracting a third dimension using only a single-lens camera. In the end, Brooks's long nights with Moravec seeded his disaffection and break with the GOFAI tradition.

As Moravec's sidekick, Brooks would also spend a good deal of time working on the Stanford Cart. In the mid-1970s, the mobile robot's image recognition system took far too long to process its surroundings for anything deserving the name "real-time." The Cart took anywhere from a quarter of an hour to four hours to compute the next stage of its assigned journey, depending on the mainframe computer load. After it processed one image, it would lurch forward a short distance and resume scanning.[2] When the robot operated outdoors, it had even greater difficulty moving by itself. It turned out that moving shadows confused the vision recognition software of the robot. The complexity in moving shadows was an entrancing discovery for Brooks. He was aware of early experiments by W. Grey Walter, a British-American neurophysiologist credited with the design of the first simple electronic autonomous robots in 1948 and 1949, intended to demonstrate how the interconnections in small collections of brain cells might cause autonomous behavior. Grey Walter had built several robotic "tortoises" that used a scanning phototube "eye" and a simple circuit controlling motors and wheels to exhibit "lifelike" movement.

While Moravec considered simple robots the baseline for his model of the evolution of artificial intelligence, Brooks wasn't convinced. In Britain in the early fifties, Grey Walter had built surprisingly intelligent robots—a species zoologically named Machina speculatrix—costing a mere handful of Brit-

ish pounds. Now more than two decades later, "A robot rely-
ing on millions of dollars of equipment did not operate nearly
as well," Brooks observed. He noticed that many U.S. developers used Moravec's sophisticated algorithms, but he wondered
what they were using them for. "Were the internal models truly
useless, or were they a down payment on better performance
in future generations of the Cart?"[3]

After receiving his Ph.D. in 1981, Brooks left McCarthy's
"logic palace" for MIT. Here, in effect, he would turn the telescope around and peer through it from the other end. Brooks
fleshed out his "bottom-up" approach to robotics in 1986. If
the computing requirements for modeling human intelligence
dwarfed the limits of human-engineered computers, he reasoned, why not build intelligent behavior as ensembles of simple behaviors that would eventually scale into more powerful
symphonies of computing in robots as well as other AI applications? He argued that if AI researchers ever wanted to realize
their goal of mimicking biological intelligence, they should start
at the lowest level by building artificial insects. The approach
precipitated a break with McCarthy and fomented a new wave
in AI: Brooks argued in favor of a design that mimicked the
simplest biological systems, rather than attempting to match
the capability of humans. Since that time the bottom-up view
has gradually come to dominate the world of artificial intelligence, ranging from Minsky's *The Society of Mind* to the more
recent work of electrical engineers such as Jeff Hawkins and
Ray Kurzweil, who both have declared that the path to human-
level AI is to be found by aggregating the simple algorithms
they see underlying cognition in the human brain.

Brooks circulated his critique in a 1990 paper titled "Elephants Don't Play Chess,"[4] arguing that mainstream symbolic
AI had failed during the previous thirty years and a new
approach was necessary. "*Nouvelle AI* relies on the emergence of more global behavior from the interaction of smaller
behavioral units. As with heuristics there is no *a priori* guar-

antee that this will always work," he wrote. "However, careful design of the simple behaviors and their interactions can often produce systems with useful and interesting emergent properties."[5]

Brooks did not win over the AI establishment overnight. At roughly the same time that he started designing his robotic insects, Red Whittaker at Carnegie Mellon was envisioning walking on the surface of Mars with Ambler, a sixteen-foot-tall six-legged robot weighing 5,500 pounds. In contrast, Brooks's Genghis robot was a hexapod weighing just over two pounds. Genghis became a poster child for the new style of AI: "fast, cheap, and out of control"—as the title read of a 1989 article that Brooks cowrote with his grad student Anita M. Flynn. Brooks and Flynn proposed that the most practical way to explore space was by sending out his low-cost insect-like robots in swarms rather than deploying a monolithic overengineered and expensive system.

Predictably, NASA was initially dismissive of the idea of building robotic explorers that were "fast, cheap, and out of control." When Brooks presented his ideas at the Jet Propulsion Laboratory, engineers who had been working on costly scientific instruments rejected the idea of a tiny inexpensive robot with limited capabilities. He was undeterred. In the late 1980s and early 1990s, Brooks's ideas resonated with the design principles underpinning the Internet. A bottom-up ideology, with components assembling themselves into more powerful and complex systems, had captured the popular imagination. With two of his students, Brooks started a company and set out to sell investors on the idea of privately sending small robots into space, first to the moon and later to Mars.[6] For $22 million, Brooks proposed, you would not only get your logo on a rover; you could also promote your company with media coverage of the launch. Movies, cartoons, toys, advertising in moondust, a theme park, and remote teleoperation—these were all part of one of the more extravagant marketing campaigns ever conceived. Brooks was aiming for a moon launch in 1990, the first one since 1978, and then planned to send another rocket to Mars just three years later. By 2010, the scheme called for sending micro-robots to Mars, Neptune, its moon Triton, and the asteroids.

What the plan lacked was a private rocket to carry the robots. The trio spoke with six private rocket launch companies, none of which at the time had made a successful launch. All Brooks needed was funding. He didn't find any investors in the private sector, so the company pitched another space organization called the Ballistic Missile Defense Organization, which was the Pentagon agency previously tasked to build the Strategic Defense Initiative, a feared and ridiculed Star Wars–style missile defense shield. The project, however, had stalled after the fall of the Soviet Union. For a while, however, the BMDO considered competing with NASA by organizing its own moon launch. The MIT trio built a convincing moon launch rover prototype christened Grendel, intended to hitchhike to the moon aboard a converted "Brilliant Pebble," the Star Wars launch vehicle originally created to destroy ICBMs by colliding with them in space. Grendel was built in accordance to Brooks's bottom-up behavior approach, and it had a successful trial, but that was as far as it got.

The Pentagon's missile division lost its turf war with NASA. The nation was unwilling to pay for two spacefaring organizations. Ironically enough, years later the developers of the NASA Sojourner, which landed on Mars in 1997, borrowed heavily from the ideas that Brooks had been proposing. Although he never made it into space, a little more than a decade later, Brooks's bottom-up approach found a commercial niche. iRobot, successor of Brooks's spacefaring company, gained success by selling an autonomous vacuum cleaner for the civilian market, while a modified mil-spec version toured the Afghanistan and Iraq terrain sniffing out improvised explosive devices.

Eventually, Brooks would win the battle with the old guard.

He found an audience for his ideas about robotics at MIT, and won accolades for what he liked to call nouvelle AI. A new generation of MIT grad students started following him and not Minsky. Nouvelle AI had a widespread impact beyond the United States and especially in Europe, where attention had shifted from the construction of human-level AIs to systems that would exhibit emergent behaviors in which more powerful or intelligent capabilities would be formed from the combination of many simpler ones.

Brooks's own interests shifted away from autonomous insects and toward social interactions with humans. With graduate students, he began designing socializing robots. Robots like Cog and Kismet, designed with graduate student Cynthia Breazeal, were used to explore human-robot interaction as well as the capabilities of the robots themselves. In 2014 Breazeal announced that she planned to commercialize a home robot growing out of that original research. She has created a plucky Siri-style family companion that remains stationary on a kitchen counter, hopefully assisting with a variety of household tasks.

In 2008, Brooks retired from the MIT AI Lab and started a low-profile company with a high-profile name, Heartland Robotics. The name evoked the problem Brooks was trying to solve: the disappearance of manufacturing from the United States as a consequence of lower overseas wages and production costs. As energy and transportation costs skyrocket, however, manufacturing robots offer a potential way to level the playing field between the United States and low-wage nations. For several years there were tantalizing rumors about what Brooks had in mind. He had been working on humanoid robots for almost a decade, but at that point the robotics industry hadn't even managed to successfully commercialize toy humanoid robots, let alone robots capable of practical applications.

When Baxter was finally unveiled in 2012, Heartland had changed its name to Rethink, with the humanoid robot receiving mixed reviews. Not everyone understood or agreed with Brooks's deliberate choice of approximating the human anatomy. Today many of his competitors sell robot arms that make no effort to mimic a human counterpart, opting for simplicity and function. Brooks, however, is undeterred. His intent is to build a robot that is ready to collaborate with rather than replace human workers. Baxter is one of a generation of robots intended to work in proximity to flesh-and-blood coworkers. The technical term for this relationship is "compliance," and there is widespread belief among roboticists that over the next half decade these machines will be widely used in manufacturing, distribution, and even retail positions. Baxter is designed to be programmed easily by nontechnical workers. To teach the robot a new repetitive task, humans only have to guide the robot's arms through the requisite motions and Baxter will automatically memorize the routine. When the robot was introduced, Rethink Robotics demonstrated Baxter's capability to

Rodney Brooks rejected early artificial intelligence in favor of a new approach he described as "fast, cheap, and out of control." Later he designed Baxter, an inexpensive manufacturing robot intended to work with, rather than replace, human workers. (*Photo courtesy of Evan McGlinn/New York Times/Redux*)

slowly pick up items on a conveyor belt and place them in new locations. This seemed like a relatively limited contribution to the workplace, but Brooks argues that the system will develop a library of capabilities over time and will increase its speed as new versions of its software become available.

It is perhaps telling that one of Rethink's early venture investors was Jeff Bezos, the chief executive of Amazon. Amazon has increasingly had problems with its nonunionized warehouse workers, who frequently complain about poor working conditions and low wages. When Amazon acquired Kiva Systems, Bezos signaled that he was intent on displacing as much human labor from his warehouses as possible. In modern consumer goods logistics there are two levels of distribution: storing and moving whole cases of goods, and retrieving individual products from those cases. The Kiva system consists of a fleet of mobile robots that are intended to save human workers the time of walking through the warehouse to gather individual products to be shipped together in a composite order. While the humans work in one location, the Kiva robots maneuver bins of individual items at just the right time in the shipping process, and the humans pick out and assemble the products to be shipped. Yet Kiva is clearly an interim solution toward the ultimate goal of building completely automated warehouses. Today's automation systems cannot yet replace human hands and eyes. The ability to quickly recognize objects among dozens of possibilities and pick them up from different positions remains a uniquely human skill. But for how long? It doesn't take much imagination to see Baxter, or a competitor from similar companies like Universal Robots or Kuka, working in an Amazon warehouse alongside teams of mobile Kiva robots. Such lights-out warehouses are clearly on the horizon, whether they are made possible by "friendly" robots like Baxter or by more impersonal systems like the ones Google's new robotics division is allegedly designing.

As the debates over technology and jobs reemerged in 2012 in the United States, many people were eager to criticize Brooks's Baxter and its humanoid design. Automation fears have ebbed and flowed in the United States for decades, but because Rethink built a robot in the human form, many thought that Rethink was building machines that should be— and now are—capable of replacing human labor. Today, Brooks argues vociferously that robots don't simply kill jobs. Instead, by lowering the cost of manufacturing in the United States, robots will contribute to rebuilding the nation's manufacturing base in new factories with jobs for more skilled, albeit perhaps fewer, workers.

The debate over humans and machines continues dogging Brooks wherever he travels. At the end of the school year in 2013, he spoke to the parents of graduating students at Brown University. The ideas in his Baxter pitch were straightforward and ones that he assumed would be palatable to an Ivy League audience. He was creating a generation of more intelligent tools for workers, he argued, and Baxter in particular was an example of the future of the factory floor, designed to be used and programmed by average workers. But the mother of one of the students would have none of it. She put up her hand and indignantly asked, "But what about jobs? Aren't these robots going to take away jobs?" Patiently, Brooks explained himself again. This was about collaborating with workers, not replacing them outright. As of 2006 the United States was sending vast sums annually to China to pay for manufacturing there. That money could provide more jobs for U.S. workers, he pointed out. You're just speaking locally, she retorted. What about globally? Brooks threw up his hands. China, he contended, needs robots even more than the United States does. Because of their demographics and particularly the one-child policy, the Chinese will soon face a shortage of manufacturing workers. The deeper point that Brooks felt he couldn't get across to a group of highly educated upper-middle-class parents was that the repetitive jobs

Baxter will destroy are not high-quality ones that should be preserved. When the Rethink engineers went into factories, they asked the workers whether they wanted their children to have similar jobs. "Not one of them said yes," he noted.

In his office, Brooks keeps photos of the factory manufacturing line of Foxconn, the world's largest contract maker of consumer electronics products. They are haunting evidence of the kinds of drudgery he wants Baxter to replace. Yet despite his obvious passion for automation and robotics, Brooks has remained more of a realist than many of his robotics and AI brethren in Silicon Valley. Although robots are indeed sprouting legs and moving around in the world among us, they are still very much machines, in Brooks's view. Despite the deeply ingrained tendency of humans to interact with robots as if they have human qualities, Brooks believes that we have a long way to go before intelligent machines can realistically match humans. "I'll know they have gotten there," he said, "when my graduate students feel bad about switching off the robot."

He likes to torment his longtime friend and MIT colleague Ray Kurzweil, who is now chartered to build a Google-scale artificially intelligent mega-machine, after having previously gained notoriety for an impassioned and detailed argument that immortality is within the reach of the current human generation through computing, AI, and extraordinary dietary supplements. "Ray, we are both going to die," he has told Kurzweil. Brooks merely hopes that a future iteration of Baxter will be refined enough to provide his elder care when the day comes.

The idea that we may be on the verge of an economy running largely without human intervention or participation (or accidents) isn't new. Almost all of the arguments in play today harken back to earlier disputes. Lee Felsenstein is the product of this eclectic mix of politics and technology. He grew up in Philadelphia, the son of a mother who was an engineer and a father who was a commercial artist employed in a loco-motive factory. Growing up in a tech-centric home was com-plicated by the fact that he was a "red diaper baby." His father was a member of the U.S. Communist Party, committed enough to the cause that he named Lee's brother Joe after Stalin.[7] However, like many children of Party members, Lee wouldn't learn that his parents were Communists until he was a young adult—he abruptly lost his summer college work-study posi-tion at Edwards Air Force Base, having failed a background investigation.

Lee's family was secularly Jewish, and books and learning were an essential part of their childhood. This would mean that bits and pieces of Jewish culture found their way into Lee's worldview. He grew up aware of the golem legend, Jewish lore that would come to influence his approach to the personal com-puting world that he in turn would help create. The idea of the golem can be dated back to the earliest days of Judaism. In the Torah, it connotes an unfinished human before God's eyes. Later it came to represent an animated humanoid creature made from inanimate matter, usually dust, clay, or mud. The golem, animated using kabbalistic methods, would became a fully living, obedient, but only partially human creation of the holy or particularly blessed. In some versions of the tale, the golem is animated by putting a parchment in its mouth, not unlike programming using paper tape. The first modern robot in literature was conceived by Czech writer Karel Čapek in his play *R. U. R. (Rossum's Universal Robots)* in 1921, and so the golem precedes it by several thousand years.

"Is there a warning for us today in this ancient fable?" won-ders R. H. MacMillan, the author of *Automation: Friend or Foe?*, a 1956 caution about the dangers of computerization of the workplace. "The perils of unrestricted 'push-button warfare' are apparent enough, but I also believe that the rapidly increas-ing part that automatic devices are playing in the peace-time industrial life of all civilized countries will in time influence their economic life in a way that is equally profound."[8]

Felsenstein's interpretation of the golem fable was perhaps more optimistic than most. Influenced by Jewish folklore and the premonitions of Norbert Wiener, he was inspired to sketch his own vision for robotics. In Felsenstein's worldview, when robots were sufficiently sophisticated, they would be neither servants nor masters, but human partners. It was a perspective in harmony with Engelbart's augmentation ideas.

Felsenstein arrived in Berkeley roughly a decade after Engelbart had studied there as a graduate student in the fifties. Felsenstein became a student during the frenetic days of the Free Speech Movement. In 1973, as the Vietnam War wound down, he set out alongside a small collective of radicals to create a computing utility that offered the power of mainframe computers to the community. They found warehouse space in San Francisco and assembled a clever computing system from a cast-off SDS 940 mainframe discarded by Engelbart's laboratory at Stanford Research Institute. To offer "*computing* power to the people," they set up free terminals in public places in Berkeley and San Francisco, allowing for anonymous access.

Governed by a decidedly socialist outlook, the group eschewed the idea of personal computing. Computing should be a social and shared experience, Community Memory decreed. It was an idea before its time. Twelve years before AOL and the Well were founded, and seven years before dial-up BBSes became popular, Community Memory's innovators built and operated bulletin boards, social media, and electronic communities from other people's cast-offs. The first version of the project lasted only until 1975 before shutting down.

Felsenstein had none of the anti-PC bias shared by both his radical friends and John McCarthy. Thus unencumbered, he became one of the pioneers of personal computing. Not only was Felsenstein one of the founding members of the Homebrew Computer Club, he also designed the Sol-20, an early hobbyist computer released in 1976, followed up in 1981 with the Osborne 1, the first mass-produced portable computer. Indeed, Felsenstein had a broad view of the impact of computing on society. He had grown up in a household where Norbert Wiener's *The Human Use of Human Beings* held a prominent place on the family bookshelf. His father had considered himself not merely a political radical but a modernist as well. Felsenstein would later write that his father, Jake, "was a modernist who believed in the perfectibility of man and the machine as the model for human society. In play with his children he would often imitate a steam locomotive in the same fashion other fathers would imitate animals."[9]

The discussion of the impact of technology had been a common trope in the Felsenstein household in the late fifties and the early sixties before Felsenstein left for college. The family discussed the impact of automation and the possibility of technological unemployment with great concern. Lee had even found and read a copy of Wiener's *God and Golem, Inc.,* published the year that Wiener had unexpectedly died while visiting Stockholm and consisting mostly of his premonitions, both dire and enthusiastic, about the consequences of machines and automation on man and society. To Felsenstein, Wiener was a personal hero.

Despite his early interest in robots and computing, Felsenstein had never been enthralled with rule-based artificial intelligence. Learning about Engelbart's intelligence amplification ideas would change the way he thought about computing. In the mid-1970s Engelbart's ideas were in the air among computer hobbyists in Silicon Valley. Felsenstein, and a host of others like him, were dreaming about what they could do with their own computers. In 1977, the second year of the personal computer era, he listened to a friend and fellow computer hobbyist, Steve Dompier, talk about the way he wanted to use a computer. Dompier described a future user interface that would be designed like a flight simulator. The user would "fly" through a computer file structure, much in the way 3-D computer programs now simulate flying over virtual terrain.

Felsenstein's thinking would follow in Dompier's footsteps. He developed the idea of "play-based" interaction. Ultimately he extended the idea to both user interface design and robotics. Traditional robotics, Felsenstein decided, would lead to machines that would displace humans, but "golemics," as he described it, using a term first introduced by Norbert Wiener, was the right relationship between human and machine. Wiener had used "golemic" to describe the pretechnological world. In his "The Golemic Approach,"[10] Felsenstein presented a design philosophy for building automated machines in which the human user was incorporated into a system with a tight feedback loop between the machine and the human. In Felsenstein's design, the human should retain a high level of skill to operate the system. It was a radically different approach compared to conventional robotics, in which human expertise was "canned" in the robot while the human remained passive.

For Felsenstein, the automobile was a good analogy for the ideal golemic device. Automobiles autonomously managed a good deal of their own functions—automatic transmission, braking, and these days advanced cruise control and lane keeping—but in the end people maintained control of their cars. The human, in NASA parlance, remained very much in the loop.

Felsenstein first published his ideas as a manifesto in the 1979 Proceedings of the West Coast Computer Faire. The computer hobbyist movement in the mid-1970s had found its home at this annual computer event, which was created and curated by a former math teacher and would-be hippie named Jim Warren. When Felsenstein articulated his ideas, the sixties had already ended, but he remained very much a utopian: "Given the application of the golemic outlook, we can look forward, I believe, to a society in which rather than bringing about the displacement of people from useful and rewarding work, machines will effect a blurring of the distinction between work and play."[11] Still, when Felsenstein wrote his essay in the late 1970s it was possible that the golem could evolve either as a collaborator or as a Frankenstein-like monster.

Although he was instrumental in elevating personal computing from its hobbyist roots into a huge industry, Felsenstein was largely forgotten until very recently. During the 1990s he had worked as a design engineer at Interval Research Corporation, and then set up a small consulting business just off California Avenue in Palo Alto, down the street from where Google's robotics division is located today. Felsenstein held on to his political ideals and worked on a variety of engineering projects ranging from hearing aids to parapsychology research tools. He was hurled back onto the national stage in 2014 when he became a target for Evgeny Morozov, the sharp-penned intellectual from Belarus who specializes in quasi-academic takedowns of Internet highfliers and exposing post-dot-com era foibles. In a *New Yorker* essay[12] aiming at what he found questionable about the generally benign and inclusive Maker Movement, Morozov zeroed in on Felsenstein's Homebrew roots and utopian ideals as expressed in a 1995 oral history. In this interview, Felsenstein described how his father had introduced him to *Tools for Conviviality* by Ivan Illich, a radical ex-priest who had been an influential voice for the political Left in the 1960s and 1970s counterculture. Felsenstein had been attracted to Illich's nondogmatic attitude toward technology, which contrasted "convivial," human-centered technologies with "industrial" ones. Illich had written largely before the microprocessor had decentralized computing and he saw computers as tools for instituting and maintaining centralized, bureaucratic control. In contrast, he had seen how radio had been introduced into Central America and rapidly became a bottom-up technology that empowered, instead of oppressed, people. Felsenstein believed the same was potentially true for computing.[13]

Morozov wanted to prove that Felsenstein and by extension

the Maker Movement that carries on his legacy are naive to believe that society could be transformed through tools alone. He wrote that "society is always in flux" and further that "the designer can't predict how various political, social, and economic systems will come to blunt, augment, or redirect the power of the tool that is being designed." The political answer, Morozov argued, should have been to transform the hacker movements into traditional political campaigns to capture transparency and democracy.

It is an impressive rant, but Morozov's proposed solution was as ineffective as the straw man he set up and sought applause for tearing down. He focused on Steve Jobs's genius in purportedly not caring whether the personal computing technology he was helping pioneer in the mid-1970s was open or not. He gave Jobs credit for seeing the computer as a powerful augmentation tool. However, Morozov entirely missed the codependency between Jobs the entrepreneur and Wozniak the designer and hacker. It might well be possible to have one without the other, but that wasn't how Apple became so successful. By focusing on the idea that Illich was only interested in simple technologies that were within the reach of nontechnical users, Morozov rigged an argument so he would win.

However, the power of "convivial" technologies, which was Illich's name for tools that are under individual control, remains a vitally important design point that is possibly even more relevant today. Evidence of this was apparent in an interaction between Felsenstein and Illich, when the radical scholar visited Berkeley in 1986. Upon meeting him, Illich mocked Felsenstein for trying to substitute communication using computers for direct communication. "Why do you want to go *deet-deet-deet* to talk to Pearl over there? Why don't you just go talk to Pearl?" Illich asked.

Felsenstein responded: "What if I didn't know that it was Pearl that I wanted to talk to?"

Illich stopped, thought, and said, "I see what you mean."

To which Felsenstein replied: "So you see, maybe a bicycle society needs a computer."

Felsenstein had convinced Illich that their communication could create community even if it was not face-to-face. Given the rapid progress in robotics, Felsenstein and Illich's insight about design and control is even more important today. In Felsenstein's world, drudgery would be the province of machines and work would be transformed into play. As he described it in the context of his proposed "Tom Swift Terminal,"[14] which was a hobbyist system that foreshadowed the first PCs, "if work is to become play, then tools must become toys."

Today, Microsoft's corporate campus is a sprawling set of interlocking walkways, buildings, sports fields, cafeterias, and parking garages dotted with fir trees. In some distinct ways it feels different from the Googleplex in Silicon Valley. There are no brightly colored bicycles, but the same cadres of young tech workers who could easily pass for college or even high school students amble around the campus.

When you approach the elevator in the lobby of Building 99, where the firm's corporate research laboratories are housed, the door senses your presence and opens automatically. It feels like *Star Trek*: Captain Kirk never pushed a button either. The intelligent elevator is the brainchild of Eric Horvitz, a senior Microsoft research scientist and director of Microsoft's Redmond Research Center. Horvitz is well known among AI researchers as one of the first generation of computer scientists to use statistical techniques to improve the performance of AI applications.

He, like many others, began with an intense interest in understanding how human minds work. He obtained a medical degree at Stanford during the 1980s, and soon immersed himself further in graduate-level neurobiology research. One night in the laboratory he was using a probe to insert a single neuron into the brain of a rat. Horvitz was thrilled. It was a dark room and he had an oscilloscope and an audio speaker. As he listened to the neuron fire, he thought to himself, "I'm finally inside. I am somewhere in the midst of vertebrate thought." At the same moment he realized that he had no idea what the firing actually suggested about the animal's thought process. Glancing over toward his laboratory bench he noticed a recently introduced Apple IIe computer with its cover slid off to the side. His heart sank. He realized that he was taking a fundamentally wrong approach. What he was doing was no different from taking the same probe and randomly sticking it inside the computer in search of an understanding of the computer's software.

He left medicine, shifting his course of study, and started taking cognitive psychology and computer science courses. He adopted Herbert Simon, the Carnegie Mellon cognitive scientist and AI pioneer, as an across-the-country mentor. He also became close to Judea Pearl, the UCLA computer science professor who had pioneered an approach to artificial intelligence breaking with the early logic- and rule-based approach, instead focusing on recognizing patterns by building nesting webs of probabilities. This approach is not conceptually far from the neural network ideas so harshly criticized by Minsky and Papert in the 1960s. As a result, during the 1980s at Stanford, Horvitz was outside the mainstream in computer science research. Many mainstream AI researchers thought his interest in probability theory was dated, a throwback to an earlier generation of "control theory" methods.

After he arrived at Microsoft Research in 1993, Horvitz was given a mandate to build a group to develop AI techniques to improve the company's commercial products. Microsoft's Office Assistant, a.k.a. Clippy, was first introduced in 1997 to help users master hard-to-use software, and it was largely a product of the work of Horvitz's group at Microsoft Research. Unfortunately, it became known as a laughingstock failure in human-computer interaction design. It was so widely reviled that Microsoft's thriller-style promotional video for Office 2010 featured Clippy's gravestone, dead in 2004 at the age of seven.[15]

The failure of Clippy offered a unique window into the internal politics at Microsoft. Horvitz's research group had pioneered the idea of an intelligent assistant, but Microsoft Research—and hence Horvitz's group—was at that point almost entirely separate from Microsoft's product development department. In 2005, after Microsoft had killed the Office Assistant technology, Steven Sinofsky, the veteran head of Office engineering, described the attitude toward the technology during program development: "The actual feature name used in the product is never what we named it during development—the Office Assistant was famously named TFC during development. The 'C' stood for clown. I will let your active imagination figure out what the TF stood for."[16] It was clear that the company's software engineers had no respect for the idea of an intelligent assistant from the outset. Because Horvitz and his group couldn't secure enough commitment from the product development group for Clippy, Clippy fell by the wayside.

The original, more general concept of the intelligent office assistant, which Horvitz's research group had described in a 1998 paper, was very different from what Microsoft later commercialized. The final shipping version of the assistant omitted software intelligence that would have prevented the assistant from constantly popping up on the screen with friendly advice. The constant intrusions drove many users to distraction and the feature was irreversibly—perhaps prematurely—rejected by Microsoft's customers. However, the company chose not to publicly explain *why* the features required to make Clippy work well were left out. A graduate student once asked Horvitz this after a public lecture and the response given was that the features had bloated Office 97 to such an extent that it would no longer fit on its intended distribution disk.[17] (Before the Internet offered feature updates, leaving something out was the only practical option.)

Such are the politics of large corporations, but Horvitz would persist. Today, a helpful personal assistant—who resides inside a computer monitor—greets visitors to his fourth-floor glass-walled corner cubicle. The monitor is perched on a cart outside his office, and the display shows the cartoon head of someone who looks just like Max Headroom, the star of the British television series about a stuttering artificial intelligence that incorporated the dying memories of Edison Carter, an earnest investigative reporter. Today Horvitz's computerized greeter can inform visitors of where he is, set up appointments, or suggest when he'll next be available. It tracks almost a dozen aspects of Horvitz's work life, including his location and how busy he is likely to be at any moment during the day.

Horvitz has remained focused on systems that augment humans. His researchers design applications that can monitor a doctor and patient or other essential conversation, offering support so as to eliminate potentially deadly misperceptions. In another application, his research team maintains a book of morbid transcripts from plane crashes to map what can go wrong between pilots and air traffic control towers. The classic and tragic example of miscommunication between pilots and air traffic control is the Tenerife Airport disaster of 1977, during which two 747 jetliners were navigating a dense fog without ground radar and collided while one was taxiing and the other was taking off, killing 583 people.[18] There is a moment in the transcript where two people attempt to speak at the same time, causing interference that renders a portion of the conversation unintelligible. One goal in the Horvitz lab is to develop ways to avoid these kinds of tragedies. When developers integrate machine learning and decision-making capabilities into AI systems, Horvitz believes that those systems will be able to reason about human conversations and then make judgments about what part of a problem people are best capable to solve and what part should be filtered through machines. The ubiquitous availability of cheap computing and the Internet has made it easier for these systems to show results and gain traction, and there are already several examples of this kind of augmentation on the market today. As early as 2005, for example, two chess amateurs used a chess-playing software program to win a match against chess experts and individual chess-playing programs.

Horvitz is continuing to deepen the human-machine interaction by researching ways to couple machine learning and computerized decision-making with human intelligence. For example, his researchers have worked closely with the designers of the crowd-sourced citizen science tool called Galaxy Zoo, harnessing armies of human Web surfers to categorize images of galaxies. Crowd-sourced labor is becoming a significant resource in scientific research: professional scientists can enlist amateurs, who often need to do little more than play elaborate games that exploit human perception, in order to help scientists map tricky problems like protein folding.[19] In a number of documented cases teams of human experts have exceeded the capability of some of the most powerful supercomputers.

By assembling ensembles of humans and machines and designating a specific research task for each group, scientists can create a powerful hybrid research team. The computers possess staggering image recognition capabilities and they can create tables of the hundreds of visual and analytic features for every galaxy currently observable by the world's telescopes. That was very inexpensive but did not yield perfect results. In the next version of the program, dubbed Galaxy Zoo 2, computers with machine-learning models would interpret the images of the galaxies in order to present accurate specimens to human classifiers, who could then catalog galaxies with much less effort than they had in the past. In yet another refinement, the system would add the ability to recognize the particular skills of different human participants and leverage them appropriately. Galaxy Zoo 2 was able to automatically categorize the problems it faced and knew which people could contribute to solving which problem most effectively.

At a TED talk in 2013, Horvitz showed the reaction of a Microsoft intern to her first encounter with his robotic greeter. He played a clip of the interaction from the point of view of the system, which tracked her face. The young woman approached the system and, when it told her that Eric was speaking with someone in his office and offered to put her on his calendar, she balked and declined the computer's offer. "Wow, this is amazing," she said under her breath, and then, anxious to end the conversation added, "Nice meeting you!" This was a good sign, Horvitz concluded, and he suggested that this type of interaction presages a world in which humans and machines are partners.

Conversational systems are gradually slipping into our daily interactions. Inevitably, the partnerships won't always develop in the way we have anticipated. In December of 2013, the movie *Her,* a love story starring Joaquin Phoenix and the voice of Scarlett Johansson, became a sensation. *Her* was a science-fiction film set in some unspecified not-far-in-the-future Southern California, and it told the story of a lonely man falling in love with his operating system. This premise seemed entirely plausible to many people who saw it. By the end of 2013 millions of people around the globe already had several years of experience with Apple's Siri, and there is a growing sense that "virtual agents" are making the transition from novelties to the mainstream.

Part of *Her* is also about the singularity, the idea that machine intelligence is accelerating at such a pace that it will eventually surpass human intelligence and become independent, rendering humans "left behind." Both *Her* and *Transcen-*

dence, another singularity-obsessed science-fiction movie introduced the following spring, are most intriguing for the way they portray human-machine relationships. In *Transcendence* the human-computer interaction moves from pleasant to dark, and eventually a superintelligent machine destroys human civilization. In *Her,* ironically, the relationship between the man and his operating system disintegrates as the computer's intelligence develops so quickly that, not satisfied even with thousands of simultaneous relationships, it transcends humanity and . . . departs.

This may be science fiction, but in the real world, this territory had become familiar to Liesl Capper almost a decade earlier. Capper, then the CEO of the Australian chatbot company My Cybertwin, was reviewing logs from a service she had created called My Perfect Girlfriend with growing horror. My Perfect Girlfriend was intended to be a familiar chatbot conversationalist that would show off the natural language technologies offered by Capper's company. However, the experiment ran amok. As she read the transcripts from the website, Capper discovered that she had, in effect, become an operator of a digital brothel.

Chatbot technology, of course, dates back to Weizenbaum's early experiments with his Eliza program. The rapid growth of computing technology threw into relief the question of the relationship between humans and machines. In *Alone Together: Why We Expect More from Technology and Less from Each Other,* MIT social scientist Sherry Turkle expresses discomfort with technologies that increase human interactions with machines at the expense of human-to-human contact. "I believe that sociable technology will always disappoint because it promises what it can't deliver," Turkle writes. "It promises friendship but can only deliver 'performances.' Do we really want to be in the business of manufacturing friends that will never be friends?"[20] Social scientists have long described this phenomenon as the false sense of community—"pseudo-gemeinschaft"—and it is not limited to human-machine interactions. For example, a banking customer might value a relationship with a bank teller, even though it exists only in the context of a commercial transaction and such a relationship might be only a courteous, shallow acquaintanceship. Turkle also felt that the relationships she saw emerging between humans and robots in MIT research laboratories were not genuine. The machines were designed to express synthetic emotions only to provoke or elucidate specific human emotional responses.

Capper would eventually see these kinds of emotional—if not overtly sexual—exchanges in the interactions customers were having with her Perfect Girlfriend chatbots. A young businesswoman who had grown up in Zimbabwe, she had previously obtained a psychology degree and launched a business franchising early childhood development centers. Capper moved to Australia just in time to face the collapse of the dot-com bubble. In Australia, she first tried her hand at search engines and developed Mooter, which personalized search results. Mooter, however, couldn't hold its own against Google's global dominance. Although her company would later go public in Australia, she left in 2005, along with her business partner, John Zakos, a bright Australian AI researcher enamored since his teenage years with the idea of building chatbots. Together they built My Cybertwin into a business selling FAQbot technology to companies like banks and insurance companies. These bots would give website users relevant answers to their frequently asked questions about products and services. It proved to be a great way for companies to inexpensively offer personalized information to their customers, saving money by avoiding customer call center staffing and telephony costs. At the time, however, the technology was not yet mature. Though the company had some initial business success, My Cybertwin also had competitors, so Capper looked for ways to expand into new markets. They tried to turn My Cybertwin into a program that created a software avatar that would interact with other people over the Internet, even while its owner was offline. It was a powerful science-fiction-laced idea that yielded only moderately positive results.

Capper has been equivocal and remains uncommitted about whether virtual assistants will take away human jobs. In interviews, she would note that virtual assistants don't directly displace workers and would focus instead on mundane work her Cybertwins do for many companies, which she argued freed up humans to do more complex and ultimately more satisfying work. At the same time, Zakos attended conferences, making assertions that when companies ran A-B testing that compared the way the Cybertwins responded to text-based questions to the way humans in call centers responded to text-based questions, the Cybertwins outperformed the humans in customer satisfaction. They boasted that when they deployed a commercial system on the website of National Australia Bank, the country's largest bank, more than 90 percent of visitors to the site believed that they were interacting with a human rather than a program. In order to be convincing, conversational software on a bank website might need to answer about 150,000 different questions—a capability that is now easily within the range of computing and storage systems.

Despite their unwillingness to confront the human job-displacement question, the consequences of Capper and Zakos's work are likely to be dramatic. Much of the growth of the U.S. white-collar workforce after World War II was driven by the rapid spread of communications networks: telemarketers, telephone operators, and technical and sales support jobs all involved giving companies the infrastructure to connect customers with employees. Computerization transformed these occupations: call centers moved overseas and the first generation of automated switchboards replaced a good number of switchboard and telephone operators. Software companies like Nuance, the SRI spin-off that offers speaker-independent voice recognition, have begun to radically transform customer call centers and airline reservation systems. Despite consumers' rejection of "voicemail hell," system technology like My Cybertwin and Nuance will soon put at risk jobs that involve interacting with customers via the telephone. The My Cybertwin conversational technology might not be good enough to pass a full-on Turing test, but it was a step ahead of most of the chatbots that were available via the Internet at the time.

Capper believes deeply that we will soon live in a world in which virtual robots are routine human companions. She holds none of the philosophical reservations that plagued researchers like Weizenbaum and Turkle. She also had no problem conceptualizing the relationship between a human and a Cybertwin as a master-slave relationship.[21] In 2007 she began to experiment with programs called My Perfect Boyfriend and My Perfect Girlfriend. Not surprisingly, there was substantially more traffic on the Girlfriend site, so she set up a paywall for premium parts of the service. Sure enough, 4 percent of the people—presumably mostly men—who had previously visited the site were willing to pay for the privilege of creating an online relationship. These people were told that there was nothing remotely human on the other end of the connection and that they were interacting with an algorithm that could only mimic a human partner. Indeed, they were willing to pay for this service, even though already at the time there was no shortage of "sex chat" websites with actual humans on the other end of the conversation.

Maybe that was the explanation. Early in the personal computer era, there was a successful text-adventure game publisher called Infocom whose marketing slogan was: "The best graphics are in your head." Perhaps the freedom of interacting with a robot relaxed the mind precisely because there was no messy human at the other end of the line. Maybe it wasn't about a human relationship at all, but more about having control and being the master. Or, perhaps, the slave.

Whatever the psychology underpinning the interactions, it

freaked Capper out. She was seeing more of the human psyche than she had bargained for. And so, despite the fact that she had stumbled onto a nascent business, she backed away and shut down My Perfect Girlfriend in 2014. There must be a better way of building a business, she decided. It would turn out that Capper's business sense was well timed. Apple's embrace of Siri had transformed the market for virtual agents. The computing world no longer understood conversational systems as quirky novelties, but rather as a legitimate mainstream form of computer interaction. Before My Perfect Girlfriend, Capper had realized that her business must expand to the United States if it was to succeed. She raised enough money, changed the company's name from My Cybertwin to Cognea, and set up shop in both Silicon Valley and New York. In the spring of 2014, she sold her company to IBM. The giant computer firm followed its 1997 victory in chess over Garry Kasparov with a comparable publicity stunt in which one of its robots competed against two of the best human players of the TV quiz show *Jeopardy!* In 2011, the IBM Watson system triumphed over Brad Rutter and Ken Jennings. Many thought the win was evidence that AI technologies had exceeded human capabilities. The reality, however, was more nuanced. The human contestants could occasionally anticipate the brief window of time in which they could press the button and buzz in before Watson. In practice, Watson had an overwhelming mechanical advantage that had little to do with artificial intelligence. When it had a certain statistical confidence that it had the correct answer, Watson was able to press the button with unerring precision, timing its button press with much greater accuracy than its human competitors, literally giving the machine a winning hand.

The irony with regards to Watson's ascendance is that IBM has historically portrayed itself as an augmentation company rather than a company that sought to replace humans. Going all the way back to the 1950s, when it terminated its first formal foray into AI research, IBM has been unwilling to advertise that the computers it sells often displace human workers.[22] In the wake of its Watson victory, the company portrayed its achievement as a step toward augmenting human workers and stated that it planned to integrate Watson's technology into the health-care field as an intellectual aid to doctors and nurses.

However, Watson was slow to take off as a physicians' advisor, and the company has broadened its goal for the system. Today the Watson business group is developing applications that will inevitably displace human workers. Watson had originally been designed as a "question-answering" system, making progress toward the fundamental goals in artificial intelligence. With Cognea, Watson gained the ability to carry on a conversation. How will Watson be used? The choice faced by IBM and its engineers is remarkable. Watson can serve as an intelligent assistant to any number of professionals, or it can replace them. At the dawn of the field of artificial intelligence IBM backed away from the field. What will the company do in the future?

Ken Jennings, the human *Jeopardy!* champion, saw the writing on the wall: "Just as factory jobs were eliminated in the 20th century by new assembly-line robots, Brad and I were the first knowledge-industry workers put out of work by the new generation of 'thinking' machines. 'Quiz show contestant' may be the first job made redundant by Watson, but I'm sure it won't be the last."[23]

7 | TO THE RESCUE

The robot laboratory was ghostly quiet on a weekend afternoon in the fall of 2013. The design studio itself could pass for any small New England machine shop, crammed with metalworking and industrial machines. Marc Raibert, a bearded roboticist and one of the world's leading designers of walking robots, stood in front of a smallish interior room, affectionately called the "meat locker," and paused for effect. The room was a jumble of equipment, but at the far end seven imposing humanoid robots were suspended from the ceiling, as if on meat hooks. Headless and motionless, the robots were undeniably spooky. Without skin, they were cybernetic skeleton-men assembled from an admixture of steel, titanium, and aluminum. Each was illuminated by an eerie blue LED glow that revealed a computer embedded in the chest that monitored its motor control. Each of the presently removed "heads" housed another computer that monitored the body's sensor control and data acquisition. When they were fully equipped, the robots stood six feet high and weighed 330 pounds. When moving, they were not as lithe in real life as they were in videos, but they had an undeniable presence.

It was the week before DARPA would announce that it had contracted Boston Dynamics, the company that Raibert had founded two decades earlier, to build "Atlas" robots as the common platform for a new category of Grand Challenge competitions. This Challenge aimed to create a generation of mobile robots capable of operating in environments that were too risky or unsafe for humans. The company, which would be acquired by Google later that year, had already developed a global reputation for walking and running robots that were built mostly for the Pentagon.

Despite taking research dollars from the military, Raibert did not believe that his firm was doing anything like "weapons work." For much of his career, he had maintained an intense focus on one of the hardest problems in the world of artificial intelligence and robotics: building machines that moved with the ease of animals through an unstructured landscape. While artificial intelligence researchers have tried for decades to simulate human intelligence, Raibert is a master at replicating the agility and grace of human movement. He had long believed that creating dexterous machines was more difficult than many other artificial intelligence challenges. "It is as difficult to reproduce the agility of a squirrel jumping from branch to branch or a bird taking off and landing," Raibert argued, "as it is to program intelligence."

The Boston Dynamics research robots, with names like LittleDog, BigDog, and Cheetah, had sparked lively and occasionally hysterical Internet discussion about the Terminator-like quality of modern robots. In 2003 the company had received its first DARPA research contract for a biologically inspired quadruped robot. Five years later, a remarkable video on YouTube showed BigDog walking over uneven terrain, skittering on ice, and withstanding a determined kick from a human without falling. With the engine giving off a banshee-like wail, it did not take much to imagine being chased through the woods by such a contraption. More than sixteen million people viewed the video, and the reactions were visceral. For many, BigDog exemplified generations of sinister sci-fi and Hollywood robots.

Raibert, who usually wears jeans and Hawaiian shirts, was unfazed by, and even enjoyed, his Dr. Evil image. As a rule, he would shy away from engaging directly with the media, and communicated instead through a frequent stream of ever more impressive "killer" videos. Yet he monitored the comments and felt that many of them ignored the bigger picture: mobile robots were on the cusp of becoming a routine part of the way humans interact with the world. When speaking on the record, he simply said that he believed his critics were missing the point. "Obviously, people do find it creepy," he told a British technical journal. "About a third of the 10,000 or so responses we have to the BigDog videos on YouTube are from people who are scared, who think that the robots are coming for them. But the ingredient that affects us most strongly is a sense of pride that we've been able to come so close to what makes people and animals animate, to make something so lifelike."[1] Another category of comments, he pointed out, was from viewers who feigned shock while enjoying a sci-fi-style thrill.

The DARPA Robotics Challenge (DRC) underscored the desired spectrum of possibilities for the relationship between humans and robots even more clearly than the previous Grand Challenge for driverless cars. It foreshadowed a world in which robots would partner with humans, dance with them, be their slaves, or potentially replace them entirely. In the initial DRC competition in 2013, the robots were almost completely tele-operated by a human reliant on the robot's sensor data, which was sent over a wired network connection. Boston Dynamics built Atlas robots with rudimentary motor control capabilities like walking and arm movements and made them available to competing teams, but the higher-level functions that the robots would need to complete specific tasks were to be programmed independently by the original sixteen teams. Later that fall, when Boston Dynamics delivered the robots to the DRC, and also when they actually competed in a preliminary competition held in Florida at the end of the year, the robots proved to be relatively slow and clumsy.

Hanging in the meat locker waiting to be deployed to the respective teams, however, they looked poised to spring into action with human nimbleness. On a quiet afternoon it evoked a scene from the 2004 movie *I, Robot,* where a police detective played by actor Will Smith walks, gun drawn, through a vast robot warehouse containing endless columns of frozen humanoid robots awaiting deployment. In a close-up shot, the eyes of one sinister automaton focus on the moving detective before it springs into action.

Decades earlier, when Raibert began his graduate studies at MIT, he had set out to study neurophysiology. One day he followed a professor back to the MIT AI Lab. He walked into a room where one of the researchers had a robot arm lying in pieces on the table. Raibert was captivated. From then on he wanted to be a roboticist. Several years later, as a newly minted engineer, Raibert got a job at NASA's Jet Propulsion Laboratory in Pasadena. When he arrived, he felt like a stranger in a strange land. Robots, and by extension their keepers, were definitely second-class citizens compared to the agency's stars, the astronauts. JPL had hired the brand-new MIT Ph.D. as a junior engineer into a project that proved to be stultifyingly boring.

Out of self-preservation, Raibert started following the work of Ivan Sutherland, who by 1977 was already a legend in computing. Sutherland's 1962 MIT Ph.D. thesis project "Sketchpad" had been a major step forward in graphical and interactive computing, and he and Bob Sproull codeveloped the first virtual reality head-mounted display in 1968. Sutherland went to Caltech in 1974 as founding chair of the university's new computer science department, where he was instrumental in working with physicist Carver Mead and electrical engineer Lynn Conway on a new model for designing and fabricating integrated circuits with hundreds of thousands of logic elements and memory—a 1980s advance that made possible the modern semiconductor industry.

Alongside his older brother Bert, Sutherland had actually come to robotics in high school, during the 1950s. The two boys had the good fortune to be tutored by Edmund C. Berkeley, an actuary and computing pioneer who had written *Giant Brains, or Machines That Think* in 1949. In 1950, Berkeley had designed Simon, which, although it was constructed with relays and a total memory of four two-bit numbers, could arguably be considered the first personal computer.[2] The boys modified it to do division. Under Berkeley's guidance, the Sutherland brothers worked on building a maze-solving mouselike robot and Ivan built a magnetic drum memory that was capable of storing 128 two-bit numbers for a high school science project, which got Ivan a scholarship to Carnegie Institute of Technology.

Once in college, the brothers continued to work on a "mechanical animal." They went through a number of iterations of a machine called a "beastie," which was based on dry cell batteries and transistors and was loosely patterned after Berkeley's mechanical squirrel named Squee.[3] They spent endless hours trying to program the beastie to play tag.

Decades later, as the chair of Caltech's computer science department in the 1970s, Sutherland, long diverted into computer graphics, had seemingly left robot design interests behind him. When Raibert heard Sutherland lecture, he was riveted by the professor's musings on what might soon be possible in the field. Raibert left the auditorium feeling entirely fired up. He set about breaking down the bureaucratic wall that protected the department chair by sending Sutherland several polite emails, and also leaving a message with his secretary.

His initial inquiries ignored, Raibert became irritated. He devised a plan. For the next two and a half weeks, he called Sutherland's office every day at two P.M. Each day the secretary answered and took a message. Finally a gruff Sutherland returned his call. "What do you want?" he shouted. Raibert explained that he was anxious to collaborate with Sutherland and wanted to propose some possible projects. When they finally met in 1977, Raibert had prepared three ideas and Sutherland, after listening to the concept of a one-legged walking—hopping, actually—robot, brusquely declared: "Do that one!"

Sutherland would become Raibert's first rainmaker. He took him along on a visit to DARPA (where Sutherland had worked for two years just after Licklider) and to the National Science Foundation, and they came away with a quarter million dollars in research funding to get the project started. The two worked together on early walking robots at Caltech, and several years later Sutherland persuaded Raibert to move with him to Carnegie Mellon, where they continued with research on walking machines.

Ultimately Raibert pioneered a remarkable menagerie of robots that hopped, walked, twirled, and even somersaulted. The two had adjoining offices at CMU and coauthored an article on walking machines for *Scientific American* in January 1983. Raibert would go on to set up the Leg Laboratory at CMU in 1981 and then move the laboratory to MIT while he held a faculty position there from 1986 to 1992. He left MIT to found Boston Dynamics. Another young MIT professor, Gill Pratt, would continue to work in the Leg Lab, designing walking machines and related technologies enabling robots to work safely in partnership with humans.

Raibert pioneered walking machines, but it was his CMU colleague Red Whittaker who almost single-handedly created "field robotics," machines that moved freely in the physical world. DARPA's autonomous vehicle contest had its roots in Red Whittaker's quixotic scheme to build a machine that could make its way across an entire state. The new generation of mobile walking rescue robots had their roots in the work that he did in building some of the first rescue robots three and a half decades ago.

Whittaker's career took off with the catastrophe at Three Mile Island Nuclear Generating Station on March 28, 1979. He had just received his Ph.D. when there was a partial meltdown in one of the two nuclear reactors at the site. The crisis exposed how unprepared the industry was to cope with the loss of control of a reactor's radioactive fuel. It would be a half decade before robots built by Whittaker and his students would enter the most severely damaged areas of the reactor and help with the cleanup.

Whittaker's opportunity came when two giant construction firms, having spent $1 billion, failed to get into the basement of the crippled reactor to inspect it and begin the cleanup. Whittaker sent the first CMU robot, which his team assembled in six months and dubbed "Rover," into Three Mile Island in April of 1984. It was a six-wheeled contraption outfitted with lights, and a camera tethered to its controller. It was lowered into the basement, where it traversed water, mud, and debris, successfully gathering the first images of consequences of the disaster. The robot was later modified to perform inspections and conduct sampling.[4]

The success of this robot set the tone for Whittaker's can-do style of tackling imposing problems. After years of bureaucratic delays, his first company, Redzone Robotics, supplied a robot to help with the cleanup at Chernobyl, the 1986 nuclear power plant disaster in Ukraine. By the early 1990s Whittaker was working on a Mars robot for NASA. The Mars robot was large and heavy, so it was unlikely to make the first mission. Instead, Whittaker plotted to find an equally dramatic project back on Earth. Early driverless vehicle research was beginning to show promise, so the CMU researchers started experimenting letting vehicles loose on Pittsburgh's streets. What about driving across an entire state? Whittaker thought that that idea, which he called "the Grand Traverse," would prove that robots were ready to perform in the real world and not just in the laboratory. "Give me two years and a half-dozen graduate students and we could make it happen," he boasted to the *New York Times* in 1991.[5] A decade and a half later at DARPA, Tony Tether lent credence to this idea by underwriting the first autonomous vehicle Grand Challenge.

Although the roboticists finally made rapid progress in building useful robots in the early 1990s, it was only after decades of disappointment. The technology failure at Three Mile Island initially cast a pall over the robotics industry. In the June 1980 issue of *Omni* magazine, Marvin Minsky wrote a long manifesto calling for the development of telepresence technologies—mobile robots outfitted with video cameras, displays, microphones, and speakers that allow their operator to be "present" from a remote location anywhere in the connected world. Minsky used his manifesto to rail against the shortcomings of the world of robotics:

Three Mile Island really needed telepresence. I am appalled by the nuclear industry's inability to deal with the unexpected. We all saw the absurd inflexibility of present day technology in handling the damage and making repairs to that reactor. Technicians are still waiting to conduct a thorough inspection of the damaged plant—and to absorb a year's allowable dose of radiation in just a few minutes. The cost of repair and the energy losses will be $1 billion; telepresence might have cut this expense to a few million dollars.

The big problem today is that nuclear plants are not designed for telepresence. Why? The technology is still too primitive. Furthermore, the plants aren't even designed to accommodate the installation of advanced telepresence when it becomes available. A vicious circle![6]

The absence of wireless networking connectivity was the central barrier to the development of remote-controlled robots at the time. But Minsky also focused on the failure of the robotics community to build robots with the basic human capabilities to grasp, manipulate, and maneuver. He belittled the state of the art of robotic manipulators used by nuclear facility operators, calling them "little better than pliers" and noted that they were not a match for human hands. "If people had a bit more engineering courage and tried to make these hands more like human hands, modeled on the physiology of the palm and fingers, we could make nuclear reactor plants and other hazardous facilities much safer."[7]

It was an easy criticism to make, yet when the article was reprinted three decades later in *IEEE Spectrum* in 2010, the field had made surprisingly little progress. Robotic hands like those Minsky had called for still did not exist. In 2013 Minsky bemoaned the fact that even at the 2011 Fukushima meltdowns, there wasn't yet a robot that could easily open a door in an emergency. It was also clear that he remained bitter over the fact that the research community had largely chosen the vision charted by Rod Brooks, which involved hunting for emergent complex behaviors by joining simple components.

One person who agreed with Minsky was Gill Pratt, who had taken over as director of the MIT Leg Lab after Marc Raibert. Later a professor and subsequently dean at Olin College in Needham, Massachusetts, Pratt arrived at DARPA in early 2010 as a program manager in charge of two major programs. One, the ARM program, for Autonomous Robotic Manipulation, involved building the robotic hands whose absence Minsky had noted. ARM hands were specified to possess a human-like functionality for a variety of tasks: picking up objects, grasping and controlling tools designed for humans, and operating a flashlight. A second part of ARM funded efforts to connect the human brain to robotic limbs, which would give wounded soldiers and the disabled—amputees, paraplegics, and quadriplegics—new freedoms. A parallel project to ARM, called Synapse, focused on developing biologically inspired computers that could better translate a machine's perception into robotic actions.

Pratt represented a new wave at DARPA, arriving shortly after the Obama administration had replaced Tony Tether with Regina Dugan as the agency's director. Tether had moved DARPA away from its historically close relationship with academia by shifting funding to classified military contractors. Dugan and Pratt tried to repair the damage by quickly reestablishing closer relations with university campuses. Pratt's research before arriving at DARPA had focused on building robots that could navigate the world outside of the lab. The challenge was giving the robots practical control over the relatively gentle forces that they would encounter in the physical world. The best way to do this, he found, was to insert some elastic material between the robot's components and the gear train that drives them. The approach tried to mimic the function played by biological tendons located between a muscle and a joint. The springy tendon material can stretch and, when measured, indicates how much force was being applied to it. Until then the direct mechanical connection between the components that made up the arms and legs used by robots gave them both power and precision that was too inflexible—and potentially dangerous—for navigating the unpredictable physical world populated by vulnerable and litigious humans.

Pratt had not initially considered human-robot collaboration. Instead, he was interested in how the elderly safely move about in the world. Typically, the infirm used walkers and wheelchairs. As he explored the contact between humans and the tools they use, he realized that elasticity offered the humans a measure of protection against unyielding obstacles. More elastic robots, Pratt concluded, could make it possible for humans to work close to the machines without fear of being injured.

They were working with Cog, an early humanoid robot designed by Rodney Brooks's robot laboratory during the 1990s. A graduate student, Matt Williamson, was testing the robot's arm. A bug in the code caused the arm to repeatedly slap the test fixture. When Brooks inserted himself between the robot and the test bench, he became the first human to ever be spanked by a robot. It was a gentle whipping and— fortunately for his graduate students—Brooks survived. Pratt's research was an advance both in biomimicry and human-robot collaboration. Brooks adopted "elastic actuation" as a central means of making robots safe for people to work with.

When Pratt arrived at DARPA he was keenly aware that despite decades of research, most robots were still kept inside labs, not just for human safety but also to protect the robot's software from an uncontrolled environment. He had been at DARPA for a little more than a year when on March 12, 2011, the tsunami struck the Fukushima Daiichi Nuclear Power Plant. The workers inside the plant had been able to control the emergency for a short period, but then high radiation leakage forced them to flee before they could oversee a safe shutdown of the reactors. DARPA became peripherally involved in the crisis because humanitarian assistance and disaster relief is a Pentagon responsibility. (The agency tried to help out in the wake of the 9/11 attacks by sending robots to search for survivors at the World Trade Center.) DARPA officials coordinated a response at Fukushima by contacting U.S. companies who had provided assistance at Three Mile Island and Chernobyl. A small armada of U.S. robots was sent to Japan in an effort to get into the plant and make repairs, but by the time power plant personnel were trained to use them it was too late to avoid the worst damage. This was particularly frustrating because Pratt could see that a swift deployment of robots would almost certainly have been helpful and limited the damage. "The best the robots could do was help survey the extensive damage that had already occurred and take radiation readings; the golden hours for early intervention to mitigate the extent of the disaster had long since passed," he wrote.[8]

The failure led to the idea of the DARPA Robotics Challenge, which was announced in April 2012. By sponsoring a grand challenge on the scale of Tether's autonomous vehicle contest, Pratt sought to spark innovations in the robotics community that would facilitate the development of autonomous machines that could operate in environments that were hostile for humans. Teams would build and program a robot to perform a range of eight tasks[9] that might be expected in a power plant emergency, but most of them would not build the robots from scratch: Pratt had contracted with Boston Dynamics to supply Atlas humanoid robots as a joint platform to jump-start the competition.

In the dark it is possible to make out the blue glow of an unblinking eye staring into the evening gloom. This light is a retina scanner that uses the eye as a digital fingerprint. These pricey electronic sentinels are not yet commonplace, but they do show up in certain ultra–high security locations. Passing beneath their gaze is a bit like passing before the unblinking eye of some cybernetic Cerberus. The scanner isn't the only bit of info-security decor. The home itself is a garden of robotic delights. Inside in the foyer, a robotic arm gripping a mallet strikes a large gong to signal a new arrival. There are wheeled, flying, crawling, and walking machines everywhere. To a visitor, it feels like the scene in the movie *Blade Runner* in which detective Rick Deckard arrives at the home of the gene-hacker

J. F. Sebastian and finds himself in a menagerie of grotesque, quirky synthetic creatures.

The real-life J.F. lording over this lair is Andy Rubin, a former Apple engineer who in 2005 joined Google to jumpstart the company's smartphone business. At the time the world thought of Google as an unstoppable company, since it had rapidly become one of the globe's dominant computing technology companies. Inside Google, however, the company's founders were deeply concerned that their advantage in Web search, and thus their newly gained monopoly, might be threatened by the rapid shift away from desktop to handheld mobile computers. The era of desktop computing was giving way to a generation of more intimate machines in what would soon come to be known as the post-PC era. The Google founders were fearful that if Microsoft was able to replicate its desktop monopoly in the emerging world of phones, they would be locked out and would lose their search monopoly. Apple had not yet introduced the iPhone, so they had no way of knowing how fundamentally threatened Microsoft's desktop stranglehold would soon be.

In an effort to get ahead, Google acquired Rubin's small start-up firm to build its own handheld software operating system as a defense against Microsoft. Google unveiled Android in November 2007, ten months after the iPhone first appeared. During the next half decade, Rubin enjoyed incredible success, displacing not just Microsoft, but Apple, Blackberry, and Palm Computing as well. His strategy was to build an open-source operating system and offer it freely to the companies who had once paid hefty licenses to Microsoft for Windows. Microsoft found it impossible to compete with free. By 2013 Google's software would dominate the world of mobile phones in terms of market share.

Early in his career, Rubin had worked at Apple Computer as a manufacturing engineer after a stint at Zeiss in Europe programming robots. He left Apple several years later with an

Andy Rubin went on a buying spree for Google when the company decided to develop next-generation robotics technologies. Despite planning a decade-long effort, he walked away after just a year. (*Photo courtesy of Jim Wilson*/New York Times/*Redux*)

elite group of engineers and programmers to build one of the early handheld computers at General Magic. General Magic's efforts to seed the convergence of personal information, computing, and telephony became an influential and high-profile failure in the new mobile computing world.

In 1999, Rubin started Palo Alto–based Danger, Inc., a smartphone handset maker, with two close friends who had also been Apple engineers. The company name reflected Rubin's early obsession with robots. (In the 1960s science-fiction television series *Lost in Space,* a robot guardian for a young boy would say "Danger, Will Robinson!" whenever trouble loomed.) Danger created an early smartphone called the Sidekick, which was released in 2002. It attracted a diverse cult following with its switchblade-style slide-out keyboard, downloadable software, email, and backups of personal information in "the cloud." While most businesspeople were still chained to their BlackBerrys, the Sidekick found popularity among young people and hipsters, many of whom switched from PalmPilots.

Rubin was a member of a unique "Band of Brothers" who passed through Apple Computer in the 1980s, a generation of young computer engineers who came of age in Silicon Valley as disciples of Steve Jobs. Captivated by Jobs's charisma and his dedication to using good design and computing technology as levers to "change the world," they set out independently on their own technology quests. The Band of Brothers reflected the tremendous influence Jobs's Macintosh project had on an entire Silicon Valley generation, and many stayed friends for years afterward. Silicon Valley's best and brightest believed deeply in bringing the Next Big Thing to millions of people.

Rubin's robot obsession, however, was extraordinary, even by the standards of his technology-obsessed engineering friends. While working on phones at Google, he bought an $80,000 robot arm and brought it to work, determined to program it to make espresso—a project that stalled for more than a year because one step in the process required more strength than the arm could exert.

Early on, Rubin had acquired the Internet domain name android.com, and friends would teasingly even refer to him as "the android." In his home in the hills near Palo Alto, evidence of the coming world of robots was everywhere, because, once again, Andy Rubin had seen something that hadn't yet dawned on most others in Silicon Valley. Rubin would soon get the opportunity to make the case for the coming age of mobile robots on a much larger stage.

In the spring of 2013, Google CEO Larry Page received a curious email. Seated in his office at the company's Mountain View headquarters, he read a message that warned him an alien attack was under way. Immediately after he read the message, two large men burst into his office and instructed him that it was essential he immediately accompany them to an undisclosed location in Woodside, the elite community populated by Silicon Valley's technology executives and venture capitalists.

This was Page's surprise fortieth birthday party, orchestrated by his wife, Lucy Southworth, a Stanford bioinformatics Ph.D. A crowd of 150 people in appropriate alien-themed costumes had gathered, including Google cofounder Sergey Brin, who wore a dress. In the basement of the sprawling mansion where the party was held, a robot arm grabbed small boxes one at a time and gaily tossed the souvenirs to an appreciative crowd. The robot itself consisted of a standard Japanese-made industrial robot arm outfitted with a suction gripper hand driven by a noisy air compressor. It helped that the robot could "see" the party favors it was picking up. For eyes—actually a single "eye"—the robot used the same sensor Microsoft originally added to the Xbox to capture the gestures of video game players in the living room.

The box-throwing robot was a prototype designed by Industrial Perception, Inc., a small team then located in a garage just across the freeway from the Googleplex in Palo Alto. When the robot, which had already briefly become an Internet sensation after a video showing its box-tossing antics had appeared on YouTube,[10] wasn't slinging boxes, it was being prototyped as a new class of intelligent industrial labor that might take over tasks as diverse as loading and unloading trucks, packing in warehouses, working on assembly lines, and restocking grocery shelves.

Equipping the robots to understand what they are seeing was only part of the challenge. Recognizing six-sided boxes had proven not to be an insurmountable problem, although AI researchers only recently solved it. Identifying wanted items on grocery shelves, for example, is an immensely more complicated challenge, and today it still exceeds the capability of the best robot programmers. However, at the Page party, the

Yaskawa robot had no apparent difficulty finding the party favor boxes, each of which contained a commemorative T-shirt. Ironically, humans had packed each of those boxes, because the robot was not yet able to handle loose shirts.

The Industrial Perception arm wasn't the only intelligent machine at the party. A telepresence robot was out on the dance floor, swaying to the music. It was midnight in Woodside, but Dean Kamen, the inventor of the Segway, was controlling the robot from New Hampshire—where it was now three A.M.

This robot, dubbed a "Beam," was from Suitable Technologies, another small start-up just a couple of blocks away from Industrial Perception. Both companies were spin-offs from Willow Garage, a robotics laboratory funded by Scott Hassan, a Stanford graduate school classmate and friend of Page's. Hassan had been the original programmer of the Google search engine while it was still a Stanford research project. Willow Garage was his effort to build a humanoid robot as a research platform. The company had developed a freely available operating system for robotics as well as a humanoid telepresence robot, PR2, that was being used in a number of universities.

That evening, both AI and IA technologies were thus in attendance at Page's party—one of the robots attempted to replace humans while another attempted to augment them. Later that year Google acquired Industrial Perception, the box-handling company, for Rubin's new robot empire.

Scott Hassan's Willow Garage spin-offs once again pose the "end of work" question. Are Page and Hassan architects of a generation of technology that will deeply disrupt the economy by displacing both white-collar and blue-collar workers? Viewed as a one-to-one replacement for humans, the Industrial Perception box handler, which will load or unload a truck, is a significant step into what has been one of the last bastions of unskilled human labor. Warehouse workers, longshoremen, and lumpers all have rough jobs that are low paying and unrewarding. Human workers moving boxes—which can weigh fifty pounds or more—roughly every six seconds get tired and often hurt their backs and wind up disabled.

The Industrial Perception engineers determined that to win contracts in warehouse and logistics operations, they needed to demonstrate that their robots could reliably move boxes at four-second intervals. Even before their acquisition by Google, they were very close to that goal. However, from the point of view of American workers, a very different picture emerges. In fact, the FedExes, UPSes, Walmarts, and U.S. Post Offices that now employ many of the nation's unskilled laborers are no longer primarily worried about labor costs and are not anxious to displace workers with lower-cost machines. Many of the workers, it turns out, have already been displaced. The companies are instead faced with an aging workforce and the reality of a labor scarcity. In the very narrow case of loading and unloading trucks, at least, it's possible the robots have arrived just in time. The deeper and as yet unanswered question remains whether our society will commit to helping its human workers across the new automation divide.

At the end of 2013 in a nondescript warehouse set behind a furniture store in North Miami, a group of young Japanese engineers began running practice sessions fully a month before the DARPA Robotics Challenge. They had studied under Masayuki Inaba, the well-known roboticist who himself was the prize student of the dean of Japanese robotics, Hirochika Inoue. Inoue had started his work in robotics during graduate school in 1965 when his graduate thesis advisor proposed that he design a mechanical hand to turn a crank.

Robots have resonated culturally in Japan more positively than they have in the United States. America has long been torn between the robot as a heroic "man of steel" and the image of a Terminator. (Of course, one might reasonably wonder what Americans really felt about the Terminator after Californians twice elected the Hollywood actor who portrayed it as governor!) In Japan, however, during the 1950s and 1960s the cartoon robot character Mighty Atom, called Astro Boy in other countries, had framed the more universally positive view of robotics. To some extent, this makes sense: Japan is an aging society, and the Japanese believe they will need autonomous machines to care for their elderly.

The Japanese team, which named themselves Schaft, came out of JSK, the laboratory Dr. Inoue had established at Tokyo University, early in 2013 with the aim of entering the DARPA Robotics Challenge. They had been forced to spin off from Tokyo University because the school, influenced by the anti-militarist period after the end of World War II, prevented the university laboratory from participating in any event that was sponsored by the U.S. military.[11] The team took its name from a 1990s Japanese musical group of the electro-industrial rock genre. Rubin had found the researchers through Marc Raibert.

When news broke that Google had acquired Schaft, it touched off a good deal of hand-wringing in Japan. There was great pride in the country's robotics technology. Not only had the Japanese excelled at building walking machines, but for years they had commercialized some of the most sophisticated robots, even as consumer products. Sony had introduced Aibo, a robotic pet dog, in 1999, and continued to offer improved versions until 2005. Following Aibo, a two-foot-tall robot, Qrio, was developed and marketed but never sold commercially. Now it appeared that Google was waltzing in and skimming the cream from decades of Japanese research.

The reality, however, is that while the Japanese dominated the first-generation robot arms, other nations are now rapidly catching up. Most of the software-centric next-generation robot development work and related artificial intelligence research was happening in the United States. Both Silicon Valley and Route 128 around Boston had once again become hotbeds of robotics start-up activity in 2012 and 2013.

When they agreed to join Rubin's expanding robot empire, the Schaft researchers felt conflicted. They expected that now that they were marching to Google's drumbeat, they would have to give up their dream of competing in the Pentagon contest. "No way!" Rubin told them. "Of course you're going to stay in the competition." The ink was barely dry on the Google contract when the Japanese engineers threw themselves into the contest. They started building three prototype machines immediately and they built mockups of each of the eight contest tasks—rough terrain, a door to open, a valve to close, a ladder, and so on—so they could start testing their robots immediately. In June, when DARPA officials checked on the progress of each group, Team Schaft's thorough preparation stunned Gill Pratt—at the time, none of the other teams had even started!

In September, when two members of the Schaft team traveled to a DARPA evaluation meeting held in Atlanta alongside the Humanoids 2013 technical conference, they brought a video to demonstrate their progress. Though they spoke almost no English, the video hit like a thunderbolt. The video showed that the young Japanese had solved all the programming problems while the other competitors were still learning how to program their robots. The other teams at the conference were visibly in shock when the two young engineers left the stage. Two months later in their Miami warehouse, the team had settled in and recreated a test course made from plywood. Even though it was almost December, muggy and miserable Miami weather and mosquitoes plagued the researchers. A local security guard who watched over the team was so bitten that he ended up in the hospital after a severe allergic reaction.

Schaft established a control station on a long table in the cavernous building. Controlling the robot was dead simple—users operated the machine with a Sony Playstation PS3 controller, just like a video game. The robot pilot borrowed sound bites from Nintendo games and added his own special audio feedback to the robot. The researchers practiced each of the tasks over and over until the robot could maneuver the course perfectly.

Homestead-Miami Speedway was no stranger to growling machines. When it hosts NASCAR races, the stands are usually filled with good ol' Southern boys. In December of 2013, however, the Robot Challenge had a decidedly different flavor. Raibert called it "Woodstock for robots." He was there to oversee both the supporting role that Boston Dynamics was playing in technical care and feeding for the Atlas humanoid robots and the splashy demonstrations of several Pentagon-funded four-legged running and walking robots. These machines would periodically trot or gallop along the racecourse to the amazement of the audience of several thousand. DARPA also hosted a robot fair with several dozen exhibitors during the two days of robot competition, which generated a modest crowd as well as a fairly hefty media contingent.

Google underscored the growing impact of robotics on all aspects of society when it publicly announced Rubin's robotics division just weeks before the Robotics Challenge. At the beginning of that month, *60 Minutes* had aired a segment about Jeff Bezos and Amazon that included a scene in which Bezos led Charlie Rose into a laboratory and showed off an octocopter drone designed to deliver Amazon products autonomously "in 30 minutes."[12] The report sparked another flurry of discussions about the growing role of robots in society. The storage and distribution of commercial goods is already a vast business in the United States, and Amazon has quickly become a dominant low-cost competitor. Google is intent on competing against Amazon in the distribution of all kinds of goods, which will create pressure to automate warehouse processes and move distribution points closer to consumers. If the warehouse was close enough to a consumer—within just blocks, for example, in a large city—why not use a drone for the "last mile"? The idea felt like science fiction come to life, and Rose, who appeared stunned, did not ask hard questions.

Google, however, unveiled its own drone delivery research project. Just days after the Amazon *60 Minutes* extravaganza, the *New York Times* reported on Google's robotic ambitions, which dwarfed what Bezos had sketched on the TV news show. Rubin had stepped down as head of Google's Android phone division in the spring of 2013. Despite reports that he had lost a power struggle and was held in disfavor, exactly the opposite was true. Larry Page, Google's chief executive, had opened the corporate checkbook and sent Rubin on a remarkable shopping spree. Rubin had spent hundreds of millions of dollars recruiting the best robotics talent and buying the best robotic technology in the world. In addition to Schaft, Google had also acquired Industrial Perception, Meka Robotics, and Redwood Robotics, a group of developers of humanoid robots and robot arms in San Francisco led by one of Rodney Brooks's star students, and Bot & Dolly, a developer of robotic camera systems that had been used to create special effects in the movie *Gravity*. Boston Dynamics was the exclamation mark in the buying spree.

Google's acquisition of an R & D company closely linked to the military instigated a round of speculation. Many suggested that Google, having bought a military robotics firm, might become a weapons maker. Nothing could have been further from the truth. In his discussions with the technologists at the companies he was acquiring, Rubin sketched out a vision of robots that would safely complete tasks performed by delivery workers at UPS and FedEx. If Bezos could dream of delivering consumer goods from the air, then how outlandish would it be for a Google Shopping Express truck to pull up to a home and dispatch a Google robot to your front door? Rubin also had long had a close relationship with Terry Gou, the CEO of Foxconn, the giant Chinese manufacturer. It would not be out of the realm of possibility to supply robots to replace some of Gou's one million "animals."

Google's timing in unveiling its new robotics effort was perfect. The December 2013 Robotics Challenge was a preliminary trial to be followed by a final event in June of 2015. DARPA organized the first contest into "tracks" broken broadly into teams that supplied their own robots and teams that used the DARPA-supplied Atlas robots from Boston Dynamics. The preliminary trial turned out to be a showcase for Google's new robot campaign. Rubin and a small entourage flew into an airport north of Miami on one of Google's G5 corporate jets and were met by two air-conditioned buses rented for the joint operation.

The contest consisted of the eight separate tasks performed over two days. The Atlas teams had a comparatively short amount of time before the event to program their robots and practice, and it showed. Compared with the nimble four-legged Boston Dynamics demonstration robots, the contestants themselves were slow and painstaking. A further reminder of how little progress had been made was that the robots were tethered from above in order to protect them from damaging falls without hampering their movements.

If that wasn't enough, DARPA gave the teams a little break in the driving task: they allowed human assistants to place the robots in

The Boston Dynamics Atlas Robot, designed for the DARPA Robot Challenge. Boston Dynamics was later acquired by Google and has designed a second-generation Atlas intended to operate without tether or power connection. (*Photo courtesy of DARPA*)

the cars and connect them to the steering wheel and brakes before they drove through a short obstacle course. Even the best teams, including Schaft, drove the course in a stop-and-go fashion, pulling forward a distance and pausing to recalibrate. The slow pace was strikingly reminiscent of the SRI Shakey robot many decades earlier. The robots were not yet autonomous. Their human controllers were hidden away in the garages while the robots performed their tasks on the speedway's infield, directed via a fiber-optic network that fed video and sensor data back to the operator console worksta-tions. To bedevil the teams and create a real-world sense of cri-sis, DARPA throttled the data connection at regular intervals. This gave even the best robots a stuttering quality, and the assembled press hunted for metaphors less trite than "watch-ing grass grow" or "watching paint dry" to describe the scene.

Nevertheless, the DARPA Robotics Challenge did what it was designed to do: expose the limits of today's robotic sys-tems. Truly autonomous robots are not yet a reality. Even the prancing and trotting Boston Dynamics machines that performed on the racetrack tarmac were wirelessly tethered to human controllers. It is equally clear, however, that truly autonomous robots will arrive soon. Just as the autonomous vehicle challenges of 2004 through 2007 significantly acceler-ated the development of self-driving cars, the Robotics Chal-lenge will bring us close to Gill Pratt's dream of a robot that can work in hazardous environments and Andy Rubin's vision of the automated Google delivery robot. What Homestead-Miami also made clear was that there are two separate paths for-ward in defining the approaching world of humans and robots, one moving toward the man-machine symbiosis that J. C. R. Licklider had espoused and another in which machines will increasingly supplant humans. Just as Norbert Wiener realized at the onset of the computer and robotics age, one of the future possibilities will be bleak for humans. The way out of that cul-de-sac will be to follow in Terry Winograd's footsteps by placing the human in the center of the design.

Darkness had just fallen on the pit lane at Homestead-Miami Speedway, giving the robotic bull trotting on the roadway a ghostlike form. The bull's machinery growled softly as its mechanical legs swung back and forth, the crate latched to its side rhythmically snapping against its trunk in a staccato rhythm. A human operator trailed the robot at a comfortable pace. Wearing a radio headset and a backpack full of communications gear, he used an oversized video game–style controller to guide the beast's pace and direction. The contraption trotted past the garages where clusters of engineers and software hackers were busy packing up robots from the day's competition.

The DRC evoked the bar scene in the Star Wars movie *Episode IV: A New Hope*. Boston Dynamics designed most of its robots in humanoid form. This was a conscious decision: a biped interacts better with man-made environments than other forms do. There were also weirder designs at the contest, like a "transformer" from Carnegie Mellon that was reminiscent of robots in Japanese sci-fi films, and a couple of spiderlike walking machines as well. The most attractive robot was Valkyrie, a NASA robot that resembled a female Star Wars Imperial Stormtrooper. Sadly, Valkyrie was one of the three underperformers in the competition; it completed none of the tasks successfully. NASA engineers had little time to refine its machinery because the shutdown of the federal government cut funds for development.

The star of the two-day event was clearly the Team Schaft robot. The designers, a crew of about a dozen Japanese engineers, had been the only team to almost perfectly complete all the tasks and so they easily won the first Robotics Challenge. Indeed, the Schaft robot had only made a single error: it tried to walk through a door that was slammed shut by the wind.

Gusts of wind had repeatedly blown the door out of the Japanese robot's grasp before it could extend its second arm to secure the door's spring closing mechanism.

While the competition took place, Rubin was busy moving his Japanese roboticists into a sprawling thirty-thousand-square-foot office perched high atop a Tokyo skyscraper. To ensure that the designers did not disturb the building's other tenants—lawyers, in this case—Google had purchased two floors in the building and decided to leave one floor as a buffer for sound isolation.

In the run-up to the Robotics Challenge, both Boston Dynamics and several of the competing teams had released videos showcasing Atlas's abilities. Most of the videos featured garden-variety demonstrations of the robot walking, balancing, or twisting in interesting ways. However, one video of a predecessor to Atlas showed the robot climbing stairs and crossing an obstacle field that involved spreading its legs across a wide gap while balancing its arms against the walls of the enclosure. It moved at human speed and with human dexterity. The video had been carefully staged and the robot was being teleoperated—it was not acting autonomously.[13] But the implications of the video were clear—the hardware for robots was capable of real-world mobility when the software and sensors caught up.

While public reaction to the video was mixed, the Schaft team loved it. In the wake of their victory, they watched in amazement as the Boston Dynamics robotic bull trotted toward their garage. It squatted on the ground and shut down. The team members swarmed around the robot and opened the crate that was strapped to its back. It contained a case of champagne, brought as a congratulatory offering from the Boston Dynamics engineers in an attempt to bond the two groups of roboticists who would soon be working together on some future Google mobile robot.

Several of the company's engineers had considered doing

something splashier. While planning for the Boston Dynamics demonstrations at the speedway, executives at another one of Rubin's AI companies came up with a PR stunt to unveil at the Boston Dynamics demonstrations during both afternoons of the Robotics Challenge. The highlight of the two-day contest had not been watching the robots as they tried to complete a set of tasks. The real crowd-pleasers were the LS3 and Wildcat four-legged robots, both of which had come out on the raceway tarmac to trot back and forth. LS3, a robotic bull-like machine without a head, growled as it moved at a determined pace. Every once in a while, a Boston Dynamics employee pushed the machine to set it off balance. The robot nimbly moved to one side and recovered quickly—as if nothing had happened. Google initially wanted to stage something more impressive. What if they could show off a robot dog chasing a robot car? That would be a real tour de force. DARPA quickly nixed the idea, however. It would have smacked of a Google promotional and the "optics" might not play well either. After all, if robots could chase each other, what else might they chase?

Team Schaft finished the champagne as quickly as they had cracked it open. It was a heady night for the young Japanese engineers. One researcher, who staggered around with a whole bottle of champagne in his hand, ended up in a hospital and with a fierce headache the next day. As the evening wound down, the implications of Schaft's win were very clear to the crowd of about three dozen robot builders who were gathered in front of the Schaft garage that evening. Rubin's new team shared a common purpose. Machines would soon routinely move among people and would inevitably assume even more of the drudgery of human work. Designing robots that could do anything from making coffee to loading trucks was well within the engineers' reach.

The Google roboticists believed passionately that, in the long run, machines woud inevitably substitute for humans. Given enough computing power and software ingenuity, it now

seemed possible that engineers could model all human qualities and capabilities, including vision, speech, perception,
manipulation, and perhaps even self-awareness. To be sure,
it was important to these designers that they operated in the
best interests of society. However, they believed that while the
short-term displacement of humans would stoke conflict, in
the long run, automation would improve the overall well-being
of humanity. This is what Rubin had set out to accomplish.
That evening he hung back from the crowd and spoke quietly with several of the engineers who were about to embark
on a new journey to introduce robots into the world. He was
at the outset of his quest but had already won a significant
wager, which was a sign of his confidence in his team. He had
bet Google CEO Larry Page his entire salary for a year that
the Schaft team would win the DARPA trials. Luckily for Page,
Rubin's annual salary was just one dollar. Like many Google
executives', his actual compensation was much, much higher.
However, a year after launching the company's robotics division, Rubin would depart the company. He had acquired a
reputation as one of the Valley's most elite technologists. But,
by his own admission, he was more interested in creating new
projects than running them. The robot kingdom he set out to
build would remain very much a work in progress after his
abrupt departure at the end of 2014.

In the weeks after Homestead, Andy Rubin made it clear
that his ultimate goal was to build a robot that could complete
each of the competitive tasks in the challenge at the push of a
button. Ultimately, it was not to be. Months later, Google would
withdraw Schaft from the finals to focus on supplying state-of-
the-art second-generation Atlas robots for other teams to use.

Today Google's robot laboratory can be found in the very
heart of Silicon Valley, on South California Ave., which divides
College Terrace, a traditional student neighborhood that was
once full of bungalows and now has grown increasingly tony,
from the Stanford Industrial Park, which might properly be

called the birthplace of the Valley. Occupying seven hundred acres of the original Leland Stanford Jr. family farm, the industrial park was the brainchild of Frederick Terman, the Stanford dean who convinced his students William Hewlett and David Packard to stay on the West Coast and start their own business, instead of following a more traditional career path and heading east to work for the electronics giants of the first half of the last century.

The Stanford Industrial Park has long since grown from a manufacturing center into a sprawling cluster of corporate campuses. Headquarters, research and development centers, law offices, and finance firms have gathered in the shadow of Stanford University. In 1970, Xerox Corp. temporarily located its Palo Alto Research Center at South California Avenue and Hanover Street, where, shortly thereafter, a small group of researchers designed the Alto computer. Smalltalk, the Alto's software, was created by another PARC group led by computer scientist Alan Kay, a student of Ivan Sutherland at Utah. Looking for a way to compete with IBM in the emerging market for office computing, Xerox had set out to build a world-class computer science lab from scratch in the Industrial Park.

More than a decade ahead of its time, the Alto was the first modern personal computer with a windows-based graphical display that included fonts and graphics, making possible on-screen pages that corresponded precisely to final printed documents (ergo WYSIWYG, pronounced "whizziwig," which stands for "what you see is what you get"). The machine was controlled by an oddly shaped rolling appendage with three buttons wired to the computer known as a mouse. For those who saw the Alto while it was still a research secret, it drove home the meaning of Engelbart's augmentation ideas. Indeed, one of those researchers was Stewart Brand, a counterculture impresario—photographer, writer, and editor—who had masterminded the *Whole Earth Catalog*. In an article for *Rolling Stone,* Brand referred to PARC as "Shy Research Center," and was later credited with coining the term "personal computer." Now, more than four decades later, the desktop personal computers of PARC are handheld and they are in the hands of much of the world's population.

Today Google's robot laboratory sits just several hundred feet from the building where the Xerox pioneers conceived of personal computing. The proximity emphasizes Andy Rubin's observation that "Computers are starting to sprout legs and move around in the environment." From William Shockley's initial plan to build an "automatic trainable robot" at the very inception of Silicon Valley to Xerox PARC and the rise of the PC and now back to Google's mobile robotics start-up, the proximity of the two laboratories underscores how the region has moved back and forth in its efforts to alternatively extend and replace humans, from AI to IA and back again.

There is no sign that identifies Google's robot laboratory. Inside the entryway, however, stands an imposing ten-foot-high steel statue—of what? It doesn't quite look like a robot. Maybe it is meant to signify some kind of alien creature. Maybe it is a replicant? The code name of Rubin's project was "Replicant," inspired, of course, by the movie *Blade Runner*. Rubin's goal was to build and commercially introduce a humanoid robot that could move around in the world: a robot that could deliver packages, work in factories, provide elder care, and generally collaborate with and potentially replace human workers. He had set out to finish what had in effect begun nearby almost a half century earlier at the Stanford Artificial Intelligence Laboratory.

The earlier research spawned by SAIL had created a generation of students like Ken Salisbury. As a young engineer Salisbury viewed himself as less of an "AI guy," and more of a "control person." He was trained in the Norbert Wiener tradition and so didn't believe that intelligent machines needed autonomy. He had been involved in automation long enough to see the shifting balance between human and machine, and preferred to keep humans in the loop. He wanted to build a robot that, for example, could shake hands with you without crushing your hand. Luckily for Salisbury, autonomy was slow to arrive. The challenge of autonomous manipulation as humans are capable of—"pick up that red rag over there"—has remained a hard problem.

Salisbury lived at the heart of the paradox described by Hans Moravec—things that are hardest for humans are easiest for machines, and vice versa. This paradox was first clarified by AI researchers in the 1980s, and Moravec had written about it in his book *Mind Children:* "It is comparatively easy to make computers exhibit adult level performance on intelligence tests or playing checkers, and difficult or impossible to give them the skills of a one-year-old when it comes to perception and mobility."[14] John McCarthy would frame the problem by challenging his students to reach into their pocket, feel a coin, and identify that coin as a nickel. Build a robot that could do that! Decades later, Rodney Brooks was still beginning his lectures and talks with the same scenario. It was something that a human could do effortlessly. Despite machines that could play chess and *Jeopardy!* and drive cars, little progress had been made in the realms of touch and perception.

Salisbury was a product of the generation of students that emerged from SAIL during its heyday in the 1970s. While he was a graduate student at Stanford, he designed the Stanford/JPL hand, an example of the first evolution of robotic manipulators from jawed mechanical grippers into more articulated devices that mimicked human hands. His thesis had been about the geometric design of a robotic hand, but Salisbury was committed to the idea of something that worked. He stayed up all night before commencement day to get a final finger moving.

He received his Ph.D. in 1982, just a year after Brooks. Both would ultimately migrate to MIT as young professors. There, Salisbury explored the science of touch because he thought it was key to a range of unsolved problems in robotics. At MIT he became friendly with Marvin Minsky and the two spent hours discussing and debating robot hands. Minsky wanted to build hands covered with sensors, but Salisbury felt durability was more important than perception, and many designs forced a trade-off between those two qualities.

While a professor at the MIT Artificial Intelligence Laboratory he worked with a student, Thomas Massie, on a handheld controller to serve as a computer interface making three-dimensional images on a computer display something that people could touch and feel. The technology effectively blurs the line between the virtual computer world and the real world. Massie—who would later become a Tea Party congressman representing Kentucky—and his wife, both mechanical engineers, turned the idea into Sensable Devices, a company that created an inexpensive haptic—or touch—control device. After taking a sabbatical year to help found both Sensable and Intuitive Surgical, a robot surgery start-up based in Silicon Valley, Salisbury returned to Stanford, where he established a robotics laboratory in 1999.

In 2007, he created the PR1, or Personal Robot One, with his students Eric Berger and Keenan Wyrobek. The machine was a largely unnoticed tour de force. It had the capabilities to leave a building, buy coffee for Salisbury, and return. The robot asked Salisbury for some money, then made its way through a series of three heavy doors. It opened each of them by pulling the handle halfway, then turning sideways so it could fit through the opening. Then it found its way to an elevator, called it, checked to make sure that no humans were inside, entered the elevator, pressed the button for the third floor, and checked to make sure the elevator had indeed gotten to the correct floor using visual cues. The robot then left the elevator, made its way to the coffee vendor, purchased coffee, and brought it back to the lab—without spilling it and before it got cold.

The PR1 looked a little like a giant coffee can with arms, motorized wheels for traction, and stereo cameras for vision. Building it cost about $300,000 over about eighteen months.

It was generally run by teleoperation except for specific pre-programmed tasks, such as fetching coffee or a beer. Capable of holding about eleven pounds in each arm, it could perform a variety of household chores. An impressive YouTube video shows the PR1 cleaning a living room. Like the Boston Dynamics Atlas, however, it was teleoperated and that particular video was sped up eight times to make it look like it moved at human speed.[15]

The PR1 project emerged from Salisbury's lab at the same time Andrew Ng, a young Stanford professor who was an expert in machine vision and statistical techniques, was working on a similar but more software-focused project, the Stanford Artificial Intelligence Robot, or STAIR. At one point Ng gave a talk describing STAIR to the Stanford Industrial Affiliates program. In the audience was Scott Hassan, the former Stanford graduate student who had done the original heavy lifting for Google as the first PageRank algorithm programmer, the basis for the company's core search engine.

It's time to build an AI robot, Ng told the group. He said his dream was to put a robot in every home. The idea resonated with Hassan. A student in computer science first at the State University of New York at Buffalo, he then entered graduate programs in computer science at both Washington University in St. Louis and Stanford, but dropped out of both programs before receiving an advanced degree. Once he was on the West Coast, he had gotten involved with Brewster Kahle's Internet Archive Project, which sought to save a copy of every Web page on the Internet.

Larry Page and Sergey Brin had given Hassan stock for programming PageRank, and Hassan also sold E-Groups, another of his information retrieval projects, to Yahoo! for almost a half-billion dollars. By then, he was a very wealthy Silicon Valley technologist looking for interesting projects.

In 2006 he backed both Ng and Salisbury and hired Salisbury's students to join Willow Garage, a laboratory he'd already created to facilitate the next generation of robotics technology—like designing driverless cars. Hassan believed that building a home robot was a more marketable and achievable goal, so he set Willow Garage to work designing a PR2 robot to develop technology that he could ultimately introduce into more commercial projects.

Sebastian Thrun had begun building a network of connections in Silicon Valley after he arrived on a sabbatical from CMU several years earlier. One of those was Gary Bradski, an expert in machine vision at Intel Labs in Santa Clara. The company was the world's largest chipmaker and had developed a manufacturing strategy called "copy exact," a way of developing next-generation manufacturing techniques to make ever-smaller chips. Intel would develop a new technology at a prototype facility and then export that process to wherever it planned to produce the denser chips in volume. It was a system that required discipline, and Bradski was a bit of a "Wild Duck"—a term that IBM originally used to describe employees who refused to fly in formation—compared to typical engineers in Intel's regimented semiconductor manufacturing culture.

A refugee from the high-flying finance world of "quants" on the East Coast, Bradski arrived at Intel in 1996 and was forced to spend a year doing boring grunt work, like developing an image-processing software library for factory automation applications. After paying his dues, he was moved to the chipmaker's research laboratory and started researching interesting projects. Bradski had grown up in Palo Alto before leaving to study physics and artificial intelligence at Berkeley and Boston University. He returned because he had been bitten by the Silicon Valley entrepreneurial bug.

For a while he wrote academic research papers about machine vision, but he soon learned that there was no direct payoff. The papers garnered respect at places like Berkeley,

Stanford, and MIT, but they didn't resonate with the rest of Silicon Valley. Besides, he realized that what was special about Intel was its deep pockets. He decided he should be exploiting them. "I should do something that has higher leverage," he thought to himself.

In his first year at Intel he met some superstar Russian software designers who worked under contract for the chipmaker, and he realized that they could be an important resource for him. At the time, the open-source software movement was incredibly popular. His background was in computer vision, and so he put two and two together and decided to create a project to build a library of open-source machine vision software tools. Taking the Linux operating system as a reference, it was obvious that when programmers worldwide have access to an extraordinary common set of tools, it makes everybody's research a lot easier. "I should give everyone that tool in vision research," he decided.

While his boss was on sabbatical he launched OpenCV, or Open Source Computer Vision, a software library that made it easier for researchers to develop vision applications using Intel hardware. Bradski was a believer in an iconoclastic operating style that was sometimes attributed to Admiral Grace Hopper and was shared by many who liked getting things done inside large organizations. "Better to seek forgiveness than to ask permission" was his motto. Eventually OpenCV contained a library of more than 2,500 algorithms including both computer vision

Gary Bradski created a popular computer vision software library and helped design robots. He would later leave robotics to work with a company seeking to build augmented reality glasses. (Photo © 2015 by Gary Bradski)

and machine-learning software. OpenCV also hosted programs that could recognize faces, identify objects, classify human motion, and so on. From his initial team of just a handful of Intel researchers, a user community grew to more than 47,000 people, and more than ten million copies of the toolset have been downloaded to date.

Realizing that he would one day leave Intel and would need a powerful toolset for his next project, Bradski developed a second agenda. OpenCV would be his calling card when he left his job at the chipmaker. Open-source software efforts were in favor inside Intel because the company wanted leverage in its difficult relationship with Microsoft. The two companies dominated the personal computing industry, but often clashed over issues of control, strategic direction, and ultimately revenue. For a while Bradksi had tremendous support inside the laboratory: at one point, he had fifteen researchers on the OpenCV project. That moment was one of the high points of his career at Intel.

Then, Intel gave him a division award and told him, "All right, now you have to move on." "What do you mean?" he responded to his managers. "This is a decadelong project." Grudgingly, he did some other things, but he covertly kept the OpenCV project going on the side. That did not sit well inside the giant semiconductor firm. One of his Russian programmers was given a performance review demerit—"improvement required"—by management because he was associated with the program.

Intel's refusal to see the value of the project left Bradski feeling disaffected. In 2001, Intel dropped its camera division, which pushed him to the edge. "More shortsighted bean counter thinking," he decided. "Of course this is low-margin silicon, but this is a loss leader, so you can eventually profit from the whole thing!" He had no idea that the mobile computing and smartphone wave was just a half decade away, but at that moment, he was right. Intel, in retrospect, had had a history of trying new ideas and then canceling them before they could bear fruit. His frustration made him an easy recruit for Sebastian Thrun, who was then building his team at Stanford to create the Stanley autonomous vehicle for the 2005 DARPA competition.

They had struck up a relationship when Thrun had been at Stanford on sabbatical in 2001. When he returned in 2003 as a faculty member, Bradski, who was disaffected with Intel, was preparing to take his own sabbatical at EPFL, a Swiss research university in Lausanne. Thrun said, "Why don't you come to Stanford instead?" Bradski was faced with a difficult decision. Switzerland would have offered him an academic feast, a chance to work on neural nets and evolutionary learning algorithms, and a great party. At the end of the day, he realized that a sabbatical at EPFL was a diversion for someone who had entrepreneurial aspirations, and the nightmare Swiss bureaucracy overwhelmed him: he should have started a year earlier getting his kids into private school, and renting a house in Lausanne was a challenge—one potential landlord told him there would be no showering after ten P.M. and he wouldn't permit noisy children!

So Bradski switched gears and took his sabbatical at relatively laid-back Stanford. He taught courses and flirted with ideas for a new start-up. His first project involved building an advanced security camera. However, he ended up with a partner who was a poor match for the project, and it quickly turned into a bad marriage. Bradski backed out. By that time, his sabbatical was over, so he went back to work at Intel and managed a large research group. He quickly realized that management involved a lot of headaches and little interesting work, so he tried to pare down his group to a core team.

Before, Bradski had been oblivious to the frustrations of other researchers, but now he noticed that engineers everywhere inside the company had similar frustrations. He joined an underground laboratory for the disaffected. Then, on a visit to Stanford, Thrun said, "Come out back to the parking lot." Thrun showed Bradski Stanley, the secret project preparing to enter the second DARPA Grand Challenge. This was obviously the coolest thing around, and Bradski immediately fell in love with the idea. Back at Intel, he quickly pulled together a secret skunkworks group to help with the computer vision system for the car. He didn't bother to ask permission. He hosted his design meetings during lunchtime and met with the Stanford team on Tuesdays.

There were immediately two problems. After Intel promised that it would not involve itself directly in the DARPA Grand Challenge, the company started sponsoring Red Whittaker's CMU team. Bradski's boss started getting complaints that Bradski was distracting people from their assigned work. "This could build up to be a firing offense," his boss told him. "We're not sponsoring the Stanford team and we're not getting into robotics." As a concession, Bradski's boss told him he could continue to work on the project personally, but could not involve other Intel Labs researchers. By then, however, Bradski no longer cared about being fired. That made everything a lot easier, and the lunchtime meetings intensified.

The tensions at Intel came to a head two days before the race. The cars and teams had arrived in Primm, Nevada, a three-casino watering hole on the California-Nevada border. Bradski called a contact in Intel's marketing department and said he needed an immediate decision about whether Intel was going to officially sponsor the Stanford car. A decal on the car would usually cost $100,000, but Thrun told him that Bradski's team had donated so much volunteer labor that they could have the sponsorship for just $20,000. The Intel marketing guy loved the idea: sponsoring two cars would double Intel's chance of backing a winner, but he balked at making an instant decision. "The money's there, but I can't just give it to you unilaterally," the executive told him.

"Look, the cars are about to be sequestered, we have half an hour left," Bradski responded.

It worked. "Okay, do it," the executive said.

Because it was so late, there was no room left on the car except a passenger window—a brilliantly visible location. Stanley won the race and Intel had backed a winner, so that was a coup. Bradski had pulled himself back from the edge of unemployment.

The vision system contributed to Stanley's success. The car relied on lasers that could sense a dynamic point cloud around the car and digital cameras that fed machine vision algorithms. In the end, the cameras saw far enough ahead that Stanley could maintain speed without slowing down. And going fast, needless to say, was necessary to win.

The glory didn't last long, however. Bradski had secured a small DARPA contract to research "cognitive architectures" with Thrun and Daphne Koller, another Stanford machine-learning expert. However, the DARPA program manager had announced his departure, which meant the grant was likely not be renewed, which in turn meant Bradski would have to look for funding elsewhere. Sure enough, Phase II was canceled as "too ambitious."

Bradski was very intrigued by robotics, so he used some of his grant money to purchase a robot arm. The $20,000 purchase set off a small explosion inside Intel's legal department. The grant money, they insisted, was restricted for hiring interns, not buying hardware, and he had to transfer the ownership of the robot arm away from Intel. Bradski gave the arm to the Stanford STAIR project, which was run by Andrew Ng. Ng was starting to explore the world of robotics with machine-learning ideas. Could they design a robot to load and unload a dishwasher? It became part of the mix leading to the PR1 robot that was brewing between Salisbury's laboratory and Ng's project.

Meanwhile, Bradski found Intel's bureaucracy more and more overbearing. He knew it was time to leave and quickly negotiated a deal to join an Israeli machine vision start-up based in San Mateo. He took the OpenCV project with him. The machine vision start-up, however, turned out to be a less than perfect match. The Israelis loved conflict and Bradski was constantly butting heads with the CTO, a former sergeant in the Israeli Army. He would usually win the arguments, but at a cost. He began job hunting again after being at the new company for just a year.

It was hard to search for a job clandestinely. He toyed with working at Facebook, who had offered him a job, but they weren't doing anything interesting in computer vision. "Come anyway," they told him. "We'll find something for you to do." To Bradksi, their recruiting seemed highly disorganized. He showed up for his interview and they told him he was late. He showed them the email that indicated that he was, in fact, on time.

"Well," they said, "you were supposed to be down the street an hour ago."

Down the street he found the building locked, closed, and dark. It occurred to him that perhaps this was some kind of weird job test, and that a camera might be following him to see what he was going to do. He kicked the door and finally someone came out. The man didn't say anything, but it seemed obvious to Bradski that he had woken him up. The guy held the door open so Bradski could go inside, then walked off silently. Bradski sat down in the dark building and before long an admin arrived and apologized for being late. There was no record of a scheduled interview and so he called the recruiter who had supposedly set everything up. After a lot of apologizing and some more runaround, Bradski had his interview with Facebook's CTO. A few days later, he had his second interview with a higher-ranking executive. The Facebook offer would have given him a lot of stock, but going to work for Facebook didn't make much sense. Miserable with the Israelis, Bradski realized he would also be miserable at Facebook, where he would most likely be forced to work on uninteresting projects. So he kept hedging. The longer he held out, the more stock Facebook offered. At that point, the job was probably worth millions of dollars, but would cause Bradski great unhappiness in what seemed like a pressure cooker.

One day, Andrew Ng called Bradski and told him he needed to meet an interesting new group of roboticists at a research lab called Willow Garage. Founded by Hassan, it was more of a research lab than a start-up. Hassan was preparing to hire seventy to eighty roboticists to throw things against the wall and see what stuck. It fit within a certain Silicon Valley tradition; labs like Xerox PARC and Willow Garage were not intended to directly create products. Rather they experimented with technologies that frequently led in unexpected directions. Xerox had created PARC in 1970 and Paul Allen had financed David Liddle to "do PARC right" when he established Interval Research in 1992. In each case the idea was to "live in the future" by building technologies that were not quite mature but soon would be. Now it looked like robotics was ripe for commercialization.

Initially Bradski was hesitant about going by for a quick lunchtime handshake. He would have to race down and back or the Israelis would notice his absence. Ng insisted. Bradski realized that Andrew was usually right about these things and so decided to give it a shot. Everything clicked. At the end of the afternoon, Bradski was still there and he no longer cared about his start-up. This was where he should be. At the end of the day, while he was still sitting in the Willow Garage parking lot, he called Facebook to say he wasn't interested. Shortly afterward, he quit his start-up.

In December of 2007 Bradski was hired to run the vision group for the next generation of Salisbury and Ng's earlier robot experiments, morphing PR1 into PR2. They built the robot and then ran it through a series of tests. They wanted the robot to do more than retrieve a beer from the fridge. They "ran" a marathon, maneuvering the robot for twenty-six miles inside the company office while Google cofounder Sergey Brin was in attendance. Afterward, they instructed the robot to find and plug itself into ten wall sockets within an hour. "Now they can escape and fend for themselves," Bradski told friends via email.

PR2 wasn't the first mobile robot to plug itself in, however. That honor went to a mobile automaton called "The Beast," designed at the Johns Hopkins Applied Physics Lab in 1960—but it could do little else.[16] PR2 was Shakey reborn half a century later. This time, however, the robot was far more dexterous. Pieter Abbeel, a University of California at Berkeley roboticist, was given one of eight PR2s that were distributed to universities. With his students, he taught the machine to fold laundry—albeit very slowly.

Though the Willow Garage team had made a great deal of progress, their research revealed to them just how far they were from developing a sophisticated machine that could function autonomously in an ordinary home. Kurt Konolige, a veteran SRI roboticist recruited by Bradski to Willow Garage, had told Bradski that these were decadelong technology development projects. They would need to refine each step dozens of times before they got everything right.

In the end, however, like Paul Allen, who had decided to pull the plug on Interval Research after just eight years of its planned ten-year life span, Scott Hassan proved not to have infinite patience. Bradski and Konolige looked on in dismay as the Willow Garage team held endless brainstorming sessions to try to come up with home robot ideas that they could commercialize relatively quickly. They both realized the lab was going to be closed. Bradski believed he knew what people really wanted in their homes—a French maid—and that wasn't going to be possible anytime soon. In his meetings with Hassan, Bradski pleaded for his team to be permitted to focus instead on manufacturing robotics, but he was shot down every time. Hassan was dead-set on the home. Eventually, Konolige didn't even bother to show up at one of the meetings—he went kayaking instead.

For a while Bradski tried to be a team player, but then he realized he was in danger of reentering the world of compromises that he had left at Intel.

"What the hell," he thought. "This isn't me. I need to do what I want."

He started thinking about potential applications for industrial robotics integration, from moving boxes to picking up products with robot arms. After discussing robotics extensively with people in industry, he confirmed that companies were hungry for robots. He told Willow's CEO that it was essential to have a plan B in case the home robot developments didn't pan out. The executive grudgingly allowed Bradski to form a small group to work on industrial applications.

Combining robot arms with new machine vision technology, Bradski's group made rapid progress, but he tried to keep word of the advances from Hassan. He knew that if word got out, the project would quickly be commercialized. He did not want to be kicked out of the Willow Garage "nest" before he was ready to launch the new venture. Finally, early in 2012, one of the programmers blabbed to the Willow Garage founder about their success and the industrial interest in robotics. Hassan sent the group an email: "I will fund this tomorrow, let's meet on Friday morning."

With Konolige and several others, and with start-up funding from Hassan, Bradski created Industrial Perception, Inc., a robotic arm company with a specific goal—loading and unloading boxes from trucks such as package delivery vehicles. After Bradski left to cofound Industrial Perception, Willow gradually disintegrated. Willow was divvied up into five companies, several robot standards efforts, and a consulting group. It had been a failure, but home robots—except for robotic vacuum cleaners—were still a distant goal.

Bradski's new company set up operations in an industrial neighborhood in South Palo Alto. The office was in a big garage, which featured one room of office cubicles and a large unfinished space where they set up stacks of boxes for the robots to endlessly load and unload. By this point, Industrial Perception had garnered interest from giant companies like Procter & Gamble, which was anxious to integrate automation technologies into its manufacturing and distribution operations. More importantly, Industrial Perception had a potential first customer: UPS, the giant package delivery firm, had a very specific application in mind—replacing human workers who loaded and unloaded their trucks.

Industrial Perception made an appearance at just one trade show, Automatica, in Chicago in January 2012. As it turned out, they didn't even need that much publicity. A year later, Andy Rubin visited their offices. He was traveling the country, scouting and acquiring robotics firms. He told those he visited that in ten to fifteen years, Google would become the world's delivery service for information *and* material goods. He needed to recruit machine vision and navigation technologies, and Industrial Perception had seamlessly integrated these technologies into their robotic arms so they could move boxes. Along with Boston Dynamics and six other companies, Rubin secretly acquired Industrial Perception. The deals, treated as "nonmaterial" by Google, would not become public for more than six months. Even when the public found out about Google's new ambitions, the company was circumspect about its plans. Just as with the Google car, the company would keep any broader visions to itself until it made sense to do otherwise.

For Rubin, however, the vision was short-lived. He tried to persuade Google to let him run his new start-up independently from what he now saw as a claustrophobic corporate culture. He lost that battle, so at the end of 2014 he left the company and moved on to create an incubator for new consumer electronics start-up ideas.

The majority of the Industrial Perception team was integrated into Google's new robotics division. Bradski, however, turned out to be too much of a Wild Duck for Google as well—which was fortuitous, because Hassan still had plans for him. He introduced Bradski to Rony Abovitz, a successful young roboticist who had recently sold Mako Surgical, a robotic surgery company that developed robots to provide support to less-experienced surgeons. Abovitz had another, potentially even bigger idea, and he needed a machine vision expert.

Abovitz believed he could reinvent personal computing so it could serve as the ultimate tool for augmenting the human mind. If he was right, it would offer a clear path to merging the divergent worlds of artificial intelligence and augmentation. At Mako, Abovitz had used a range of technologies to digitally capture the skills of the world's best surgeons and integrate them into a robotic assistant. This made it possible for a less-skilled surgeon to use a robotic template to get consistently good results using a difficult technique. The other major robot surgery company, Intuitive Surgical, was an SRI spin-off that sold teleoperated robotic instruments that allowed surgeons to operate *remotely* with great precision. Abovitz instead focused on the use of haptics—giving the robot's operators a sense of touch—to attempt to construct a synthesis of human and robot, a surgeon more skilled than a human surgeon alone. It helped that Mako focused on operations that dealt with bone instead of soft tissue surgery (which, incidentally, was the focus of Intuitive's research). Bone, a harder material, was much easier to "feel" with touch feedback. In this system, the machine and the human would each do what they were good at to create a powerful symbiosis.

It's important to note that the resulting surgeon isn't a "cyborg"—a half-man, half-machine. A bright line between the surgeon and the robot is maintained. In this case the human surgeon works with the separate aid of a robotic surgery tool. In contrast, a cyborg is a creature in which the line between human and machine becomes blurred. Abovitz believed that

"Strong" artificial intelligence—a machine with human-level intelligence—was an extremely difficult problem and would take decades to develop, if it was ever possible. From his Mako experience designing a robot to aid a surgeon, he believed the most effective way to design systems was instead to use artificial intelligence technology to enhance human powers.

After selling Mako Surgical for $1.65 billion in late 2013, Abovitz set out to pursue his broader and more powerful augmentation idea—Magic Leap, a start-up with the modest goal of replacing both televisions and personal computers with a technology known as augmented reality. In 2013, the Magic Leap system worked only in a bulky helmet. However, the company's goal was to shrink the system into a pair of glasses less obtrusive and many times more powerful than Google Glass. Instead of joining Google, Bradski went to work for Abovitz's Magic Leap.

In 2014, there was already early evidence that Abovitz had made significant headway in uniting AI and IA. It could be seen in Gerald, a half-foot-high animated creature floating in an anonymous office complex in a Miami suburb. His four arms waved gently while he hung in space and walked in circles in front of a viewer. Gerald wasn't really there. He was actually an animated projection that resembled a three-dimensional hologram. Users could watch him through transparent lenses that project what computer scientists and optical engineers describe as a "digital light field" into the eyes of a human observer. Although Gerald doesn't exist in the real world, Abovitz is trying to create an unobtrusive pair of computer-augmented glasses with which to view animations like Gerald. And it doesn't stop with imaginary creatures. In principle it is possible to project any visual object created with the technology that matches the visual acuity of the human eye. For example, as Abovitz describes the Magic Leap system, it will make it possible for someone wearing the glasses to simply gesture with their hands to create a high-resolution screen as crisp as a flat-panel television. If they are perfected, the glasses will replace not only our TVs and computers, but many of the other consumer electronics gadgets that surround us.

The glasses are based on a transparent array of tiny electronic light emitters that are installed in each lens to project the light field—and so the image—onto each retina. In practice, computer-generated light fields attempt to mimic what the human eye sees in the physical world. It is a computer-generated version of the analog light field that comprises the sum of all of the light rays that form a visual scene for the human eye. Digital light fields simulate the way light behaves in the physical world. When photons bounce off objects in the world, they act like rivers of light. The human neuro-optic system has evolved so that the lenses in our eyes adjust to match the wavelength of the natural light field and focus on objects. Watching Gerald wander in space through a prototype of the Magic Leap glasses gives a hint that in the future it will be possible to visually merge computer-generated objects with the real world. Significantly, Abovitz claims that digital light field technology holds out the promise of circumventing the limitations that have plagued stereoscopic displays for decades. Today, these displays cause motion sickness in users and they do not offer "true" depth-of-field perception.

By January of 2015 it had become clear that augmented reality was no longer a fringe idea. With great fanfare Microsoft demonstrated a similar system called HoloLens based on a competing technology. Is it possible to imagine a world where the ubiquitous LCDs of today's modern world—televisions, computer monitors, smartphone screens—simply disappear? In Hollywood, Florida, Magic Leap's demonstration suggests that workable augmented reality is much closer than we might assume. If they are correct, such an advance would also change the way we think about and experience augmentation and automation. In October 2014, Magic Leap's technology received a significant boost when Google led a $524 million investment round in the tiny start-up.

The Magic Leap prototype glasses look like ordinary glasses, save for the thin cable that runs down a user's back and connects to a small, smartphone-sized computer. These glasses don't simply represent a break with existing display technologies. The technology behind them makes extensive use of artificial intelligence, and machine vision, to remake reality. The glasses are compelling for two reasons. First, their resolution will approach the resolving power of the human eye. The best computer displays are just reaching this level of resolution. As a result, the animations and imagery will surpass those of today's best consumer video game systems. Second, they are the first indication that it is possible to seamlessly blend computer-generated imagery with physical reality. Until now, the limits of consumer computing technology have been defined by what is known as the "WIMP" graphical interface—the windows, icons, menus, and pointer of the Macintosh and Windows. The Magic Leap glasses, however, will introduce augmented reality as a way of revitalizing personal computing and, by extension, presenting new ways to augment the human mind.

In an augmented reality world, the "Web" will become the space that surrounds you. Cameras embedded in the glasses will recognize the objects in people's environments, making it possible to annotate and possibly transform them. For example, reading a book might become a three-dimensional experience: images could float over the text, hyperlinks might be animated, readers could turn pages with the movement of their eyes, and there would be no need for limits to the size of a page.

Augmented reality is also a profoundly human-centered version of computing, in line with Xerox PARC computer scientist Mark Weiser's original vision of "calm" ubiquitous computing. It will be a world in which computers "disappear" and everyday objects acquire "magical" powers. This presents a host of new and interesting ways for humans to interact with robots. The iPod and the iPhone were the first examples of this transition as a reimagining of the phonograph and the telephone. Augmented reality would also make the idea of telepresence far more compelling. Two people separated by great distance could gain the illusion of sharing the same space. This would be a radical improvement on today's videoconferencing and awkward telepresence robots like Scott Hassan's Beam, which place a human face on a mobile robot.

Gary Bradski left the world of robots to join Abovitz's effort to build what will potentially become the most intimate and powerful augmentation technology. Now he spends his days refining computer vision technologies to fundamentally remake computing in a human-centered way. Like Bill Duvall and Terry Winograd, he has made the leap from AI to IA.

8 | "ONE LAST THING"

Set on the Pacific Ocean a little more than an hour's drive south of San Francisco, Santa Cruz exudes a Northern California sensibility. The city blends the Bohemian flavor of a college town with the tech-savvy spillover from Silicon Valley just over the hill. Its proximity to the heart of the computing universe and its deep countercultural roots are distinct counterpoints to the tilt-up office and manufacturing buildings that are sprinkled north from San Jose on the other side of the mountains. Geographically and culturally, Santa Cruz is about as far away from the Homestead-Miami Speedway as you can get.

It was a foggy Saturday morning in this eclectic beach town, just months after the Boston Dynamics galloping robots stole the show at the steamy Florida racetrack. Bundled against the morning chill, Tom Gruber and his friend Rhia Gowen wandered into The 418 Project, a storefront dance studio that backs up against the river. They were among the first to arrive. Gruber is a wiry salt-and-pepper-goateed software

Tom Gruber began his career as an artificial intelligence researcher who swung from AI to work on augmenting human intelligence. He was a cofounder of the team of programmers who designed Siri, Apple's iPhone personal assistant. (*Photo © 2015 by Tom Gruber*)

designer and Gowen is a dance instructor. Before returning to the United States several years ago, she spent two decades in Japan, where she directed a Butoh dance theater company.

In Santa Cruz, Gowen teaches a style of dance known as Contact Improvisation, in which partners stay physically in touch with each other while moving in concert with a wide range of musical styles. To the untrained eye, "Contact Improv" appears to be part dance, part gymnastics, a bit of tumbling, and even part wrestling. Dancers use their bodies in a way that provides a sturdy platform for their partners, who may roll over and even bounce off them in sync with the music. The Saturday-morning session that Gruber and Gowen attended was even more eclectic: it was a morning weekend ritual for the Santa Cruz Ecstatic Dance Community. Some basic rules are spelled out at ecstaticdance.org:

1. Move however you wish;
2. No talking on the dance floor;
3. Respect yourself and one another.

There is also an etiquette that requires that partners be "sensitive" if they want to dance with someone and that offers a way out if they don't: "If you'd rather not dance with someone, or are ending a dance with someone, simply thank them by placing your hands in prayer at your heart."

The music mix that morning moved from meditative jazz to country, rock, and then to a cascade of electronic music styles. The room gradually filled with people, and the dancers each entered a personal zone. Some danced together, some traded partners, some swayed to an inner rhythm. It was free-form dance evocative of a New Age gym class.

Gruber and Gowen wove through the throng. Sometimes they were in contact, and sometimes they broke off to dance with other partners, then returned. He picked her up and bent down and let her roll across his back. It wasn't exactly "do-si-do your partner," but if the move was done well, one body formed a platform that shouldered the other partner's weight without strain. Gruber was a confident dancer and comfortable with moves that evoked a modern dance sensibility. It offered a marked contrast to the style of many of the more hippie, middle-aged Californians, who were skipping and waving in all directions against a quickening beat. The pace of the dancers ascended to a frenzy and then backed down to a mellower groove. Gradually, the dancers melted away from the dance floor. Gruber and Gowen donned their jackets and stepped out into the still-foggy morning air.

Gruber casually pulled an iPhone from his pocket and asked Siri, the software personal assistant he designed, a simple question about his next stop. On Monday he would be back in the fluorescent-lit hallways of Apple, amid endless offices overloaded with flat-panel displays. On that morning, however, he wandered in a more human-centric world, where computers had disappeared and everyday devices like phones were magical.

Apple's corporate campus is circumscribed by Infinite Loop, a balloon-shaped street set just off the Interstate 280 freeway in Cupertino. The road wraps in a protective circle around a modern cluster of six office buildings facing inward onto a grassy courtyard. It circles a corporate headquarters that reflects Apple's secretive style. The campus was built during the era in which John Sculley ran the company. When originally completed, it served as a research and development center, but as Apple scaled down after Sculley left in 1993, it became a fortress for an increasingly besieged company. When Steve Jobs returned, first as "iCEO" in 1997, there were many noticeable changes including a dramatic improvement in the cafeteria food. The fine silver that had marked the executive suite during the brief era when semiconductor chief Gilbert Amelio ran the company also disappeared.

As his health declined during a battle with pancreatic cancer in 2011, Steve Jobs came back for one last chapter at Apple. He had taken his third medical leave, but he was still the guiding force at the company. He had stopped driving and so he would come to Apple's corporate headquarters with the aid of a chauffeur. He was bone-thin and in meetings he would mention his health problems, although never directly acknowledging the battle was with cancer. He sipped 7UP, which hinted to others that he might have been struggling through chemotherapy.

The previous spring Jobs had acquired Siri, a tiny developer of a natural language software application that was designed to act as a virtual assistant, in effect a software assistant, on the iPhone. The acquisition had drawn a great deal of attention in Silicon Valley. Apple acquisitions, particularly large ones, are extremely rare. When word circulated that the firm had been acquired, possibly for more than $200 million, it sent shock waves up and down Sand Hill Road and within the burgeoning "app economy" that the iPhone had spawned. After Apple acquired Siri, the program was immediately pulled from the App Store, the iPhone service through which programs were screened and sold, and the small team of programmers who had designed Siri vanished back into

"stealth mode" inside the Cupertino campus. The larger implications of the acquisition weren't immediately obvious to many
in the Valley, but as one of his last acts as the leader of Apple,
Steve Jobs had paved the way for yet another dramatic shift in
the way humans would interact with computers. He had come
down squarely on the side of those who placed humans in control of their computing systems.

Jobs had made a vital earlier contribution to the computing world by championing the graphical desktop computing
approach as a more powerful way to operate a PC. The shift
from the command line interface of the IBM DOS era to the
desktop metaphor of the Macintosh had opened the way for the
personal computer to be broadly adopted by students, designers, and office workers—a computer for "the rest of us," in
Apple parlance. Steve Jobs's visits to PARC are the stuff of legend. With the giant copier company's blessing and a small but
lucrative Xerox investment in Apple pre-IPO, he visited several
times in 1979 and then over the next half decade created first
the Lisa and then the Macintosh.

But the PC era was already giving way to a second Xerox
PARC concept—ubiquitous computing. Mark Weiser, the PARC
computer scientist, had conceived the idea during the late
1980s. Although he had been given less credit for the insight
and the shift, Jobs had been the first to successfully translate Weiser's ideas for general consumer audiences. The iPod
and then the iPhone were truly ubiquitous computing devices.
Jobs first transformed the phonograph and then the telephone
by adding computing. "A thousand songs in your pocket" and
"something wonderful for your hand." He was the consummate
showman, and "one more thing" had become a trademark
slogan that Jobs used at product introductions, just before
announcing something "insanely great." For Jobs, however, Siri
was genuinely his "one last thing." By acquiring Siri he took his
final bow for reshaping the computing world. He bridged the
gap between Alan Kay's Dynabook and the Knowledge Naviga-

tor, the elaborate Apple promotional video imagining a virtual personal assistant. The philosophical distance between AI and IA had resulted in two separate fields that rarely spoke. Even today, in most universities artificial intelligence and human-computer interaction remain entirely separate disciplines. In a design approach that resonated with Lee Felsenstein's original golemics vision, Siri would become a software robot—equipped with a sense of humor—intended to serve as a partner, not a slave.

It was an extraordinary demand that only Steve Jobs would have considered. He directed his phone designers to take a group of unknown software developers, who had never seen any of Apple's basic operating system software, and allow them to place their code right at the heart of the iPhone. He then forced his designers to create connections to all of the iPhone's application programs from the ground up. And he ordered that it all happen in less than a year. To supplement the initial core of twenty-four people who had arrived with the Siri acquisition, the programmers borrowed and begged from various corners of Apple's software development organization. But it wasn't enough. In most technical companies a demand of this scale would be flatly rejected as impossible. Jobs simply said, "Make it happen."

Tom Gruber was a college student studying psychology in the late 1970s when he stumbled upon artificial intelligence. Wandering through his school library, he found a paper describing the work of Raj Reddy and a group of Carnegie Mellon University computer scientists who had built a speech recognition system called Hearsay-II. The program was capable of recognizing just a thousand words spoken in sentences with a 90 percent accuracy rate. One error every ten words, of course, was not usable. What struck Gruber, though, was that the Hearsay system married acoustic signal processing with more general artificial intelligence techniques. He immediately realized that the system implied a model of the brain that was required to represent human knowledge. He realized that psychologists were also modeling this process, but poorly. At that point in the 1980s, there were no PET scans or fMRI brain-imaging systems. Psychologists were studying human behavior, but not the brain itself.

Not long after reading about the Hearsay research, Gruber found the early work of Edward Feigenbaum, a Stanford University computer science professor who focused on the idea of building "expert systems" to capture human knowledge and replicate the capabilities of specialists in highly technical fields. While he was a graduate student at Carnegie Mellon working with Herbert Simon, Feigenbaum had done research in designing computer models of human memory. The Elementary Perceiver and Memorizer, or EPAM, was a psychological theory of human learning and memory that researchers could integrate into a computer program.

Feigenbaum's work inspired Gruber to think more generally about building models of the mind. At this point, however, he hadn't considered applying to graduate school. No one in his family had studied for an advanced degree and the idea wasn't on his radar. By the time he finally sent out applications, there were only a few places that would still offer him funding. Both Stanford and MIT notified Gruber that his application was about three months late for the upcoming school year, and they invited him to apply again in the future. Luckily, he was accepted by the University of Massachusetts, which at the time was home to a vibrant AI group that was researching work in robotics including how to program robotic hands. The program's academic approach to robotics explicitly melded artificial intelligence and cognitive science, which spoke perfectly to his interest in modeling the human mind.

For Gruber, AI turned out to be the fun part of computer science. It was philosophically rich and scientifically interesting, offering ideas about psychology and the function of the human mind. In his view, the rest of computer science was really just engineering. When he arrived in Massachusetts in 1981, he worked with Paul Cohen, a young computer scientist who had been a student of Feigenbaum's at Stanford and shared Gruber's interest in AI and psychology. Paul Cohen's father, Harold Cohen, was a well-known artist who had worked at the intersection of art and artificial intelligence. He had designed the computer program Aaron and used it to paint and sell artistic images. The program didn't create an artistic style, but it was capable of generating an infinite series of complex images based on parameters set by Cohen. Aaron proved to be a powerful environment for pondering philosophical questions about autonomy and creativity.

Gruber had mentioned to the computer science department chairman that he wanted to have a social impact in his career and so he was directed to a project designing systems that would allow people with severely crippling conditions like cerebral palsy to communicate. Many of those with the worst cases couldn't speak and, at the time, used a writing system called Bliss Boards that allowed them to spell words by pointing at letters. This was a painstaking and limiting process. The system that Gruber helped develop was an early version of what researchers now call "semantic autocomplete." The researchers worked with children who could understand language clearly but had difficulty speaking. They organized the interaction scheme so the system anticipated what a participant might say next. The challenge was to create a system to communicate things like "I want a hamburger for lunch."

It was a microcosm of the entire AI world at the time. There was no big data; researchers could do little more than build a small model of the child's world. After working on this project for a while, Gruber built a software program to simulate that world. He made it possible for the caregivers and parents to add sentences to the program that personalized the system for a particular child. Gruber's program was an example of what the AI community would come to call "knowledge-based systems," programs that would reason about complex problems using rules and a database of information. The idea was to create a program that would be able to act like a human expert such as a doctor, lawyer, or engineer. Gruber, however, quickly realized that acquiring this complex human knowledge would be difficult and made this problem the subject of his doctoral dissertation.

Gruber was a skilled computer hacker, and many faculty members wanted to employ him to do their grunt work. Instead, he moonlighted at Digital Equipment Corporation, the minicomputer manufacturer. He was involved in a number of projects for DEC, including the development of an early windowing system that was written in McCarthy's AI programming language Lisp. The fact that the program ran well surprised many software developers because Lisp was not intended for graphical applications where you needed blinding speed. It took Gruber a month during the summer to write the program. It was much more common for developers to write these kinds of applications in assembly language or C in order to save time, but it turned out that for Gruber, Lisp was efficient enough. To show off the power of the Lisp programming language, he built a demo of an automated "clipping service" for visitors from the NSA. The program featured an interactive interface that allowed a computer user to tailor a search, then save it in a permanent alert system that would allow the filtering of that information. The idea stuck with him, and he would reuse it years later when he founded his first company.

Focused on getting a Ph.D. and still intrigued by science-of-mind questions, he avoided going to work for the then booming DEC. Graduate school was nirvana. He rode his bike frequently in Western Massachusetts and was able to telecommute, making more than thirty dollars an hour from his home terminal. He spent his summers in Cambridge because it was a lively place to be, working in Digital's laboratory. He also became part of a small community of AI researchers who were struggling to build software systems that approximated human expertise. The group met annually in Banff. AI researchers quickly realized that some models of human reasoning defied conventional logic. For example, engineering design is made up of a set of widely divergent activities. An HVAC—heating, ventilation, and air-conditioning—system designer might closely follow a set of rules and constraints with few exceptions. In optics, precise requirements make it possible to write a program that would design the perfect glass. Then there is messy design, product design, for example, where there are no obvious right answers and a million questions about what is required and what is optional. In this case the possible set of answers is immense and there is no easy way to capture the talent of a skilled designer in software.

Gruber discovered early in his research why conventional expert systems models failed: human expertise isn't reducible to discrete ideas or practices. He had begun by building small models, like a tool for minimizing the application of pesticides on a tree farm. Separately, he worked with cardiologists to build a diagnostic system that modeled how they used their expertise. Both were efforts to capture human expertise in software. Very simple models might work, but the complexity of real-world expertise was not easily reducible to a set of rules. The doctors had spent decades practicing medicine, and Gruber soon realized that attempting to reduce what they did to "symptoms and signs" was impossible. A physician might ask patients about what kind of pain they were experiencing, order a test, and then prescribe nitroglycerin and send them home. Medicine could be both diagnostic and therapeutic. What Gruber was seeing was a higher-level strategy being played out by the human experts, far above the rote actions of what was then possible with relatively inflexible expert system programs.

He soon realized that he wasn't interested in building better expert systems. He wanted to build better tools to make it easier for people to design better expert systems. This was to become known as the "knowledge acquisition problem." In his dissertation he made the case that researchers did not need to model knowledge itself but rather strategy—that is, knowledge about what to do next—in order to build a useful expert system. At the time, expert systems broke easily, were built manually, and required experts to compile the knowledge. His goal was to design a way to automate the acquisition of this elusive "strategic knowledge."

As a graduate student his approach was within the existing AI community framework: At the outset he defined artificial intelligence conventionally, as being about understanding intelligence and performing human-level tasks. Over time, his perspective changed. Not only should AI imitate human intelligence; he came to believe it should aim to *amplify* that intelligence as well. He hadn't met Engelbart and he wasn't familiar with his ideas, but using computing to extend, rather than simulate or replace, humans would become a motivating concept in his research.

While he was still working on his dissertation he decided to make the leap to the West Coast. Stanford was the established center for artificial intelligence research and Ed Feigenbaum, then a rising star in the AI world, was working there. He had launched a project to build the world's largest expert system on "engineering knowledge," or how things like rocket ships and jet engines were designed and manufactured. Gruber's advisor Paul Cohen introduced him to Feigenbaum, who politely told him that his laboratory was on soft money and he just didn't have any slots for new employees.

"What if I raise my own money?" Gruber responded.

"Bring your own money?!"

Feigenbaum agreed, and Gruber obtained support from some of the companies he had consulted for. Before long, he was managing Feigenbaum's knowledge engineering project. In 1989, Gruber thus found himself at Stanford University during the personal computing boom and the simultaneous precipitous decline of the AI field in the second AI Winter. At Stanford, Gruber was insulated from the commercial turmoil. Once he started on Feigenbaum's project, however, he realized that he was still faced with the problem of how to acquire the knowledge necessary to simulate a human expert. It was the same stumbling block he had tried to solve in his dissertation. That realization quickly led to a second: to transition from "building" to "manufacturing" knowledge systems, developers needed standard parts. He became part of an effort to standardize languages and categories used in the development of artificial intelligence. Language must be used precisely if developers want to build systems in which many people and programs communicate. The modules would fail if they didn't have standardized definitions. The AI researchers borrowed the term "ontology," which was the philosophical term for the study of being, using it in a restricted fashion to refer to the set of concepts—events, items, or relations—that constituted knowledge in some specific area. He made the case that an ontology was a "treaty," a social agreement among people interested in sharing information or conducting commerce.

It was a technology that resonated perfectly with the then new Internet. All of a sudden a confused world of multiple languages and computer protocols were all connected in an electronic Tower of Babel. When the World Wide Web first emerged, it offered a universal mechanism for easily retrieving documents via the Internet. The Web was loosely based on the earlier work of Doug Engelbart and Ted Nelson in the 1960s, who had independently pioneered the idea of hypertext linking, making it possible to easily access information stored in computer networks. The Web rapidly became a medium for connecting anyone to anything in the 1990s, offering a Lego-like way to link information, computers, and people.

Ontologies offered a more powerful way to exchange any kind of information by combining the power of a global digital library with the ability to label information "objects." This made it possible to add semantics, or meaning, to the exchange of electronic information, effectively a step in the direction of artificial intelligence. Initially, however, ontologies were the province of a small subset of the AI community. Gruber was one of the first developers to apply engineering principles to building ontologies. Focusing on that engineering effort drew him into collaborative work with a range of other programmers, some of whom worked across campus and others a world away. He met Jay "Marty" Tenenbaum, a computer scientist who had previously led research efforts in artificial intelligence at SRI International and who at the time directed an early Silicon Valley AI lab set up by the French oil exploration giant Schlumberger. Tenenbaum had an early and broad vision about the future of electronic commerce, preceding the World Wide Web. In 1992 he founded Enterprise Integration Technologies (EIT), a pioneer in commercial Internet commerce transactions, at a time when the idea of "electronic commerce" was still largely unknown.

From an office near the site where the Valley's first chipmaker, Fairchild Semiconductor, once stood, Tenenbaum sketched out a model of "friction free" electronic commerce. He foresaw a Lego-style automated economy in which entire industries would be woven together by computer networks and software systems that automated the interchange of goods and services. Gruber's ontology work was an obvious match for Tenenbaum's commerce system because it was a system that required using a common language to connect disparate parts. Partly as a result of their collaboration, Gruber was one of the first Silicon Valley technologists to immerse himself in the World Wide Web. Developed by Tim Berners-Lee in the heart of the particle physics community in Switzerland, the Web was rapidly adopted by computer scientists. It became known to a much wider audience when it was described in the *New York Times* in December of 1993.[1]

The Internet allowed Gruber to create a small group that blossomed into a living cyber-community expressed in the exchange of electronic mail. Even though few of the participants had face-to-face contact, they were in fact a "virtual" organization. The shortcoming was that all of their communications were point-to-point and there was no single shared copy of the group electronic conversation. "Why don't I try to build a living memory of all of our exchanges?" Gruber thought. His idea was to create a public, retrievable, permanent group memory. Today, with online conferences, support systems, and Google, the idea seems trivial, but at the time it was a breakthrough. It had been at the heart of Doug Engelbart's original NLS system, but as the personal computer had emerged, much of Engelbart's broader vision had been sidelined as first Xerox PARC and then Apple and Microsoft had cherry-picked his ideas, like the mouse and hypertext, while ignoring his broader mission for an intelligence augmentation system that would facilitate small groups of knowledge workers. Gruber created a software program that automatically generated a living document of the work done by a group of people. Over a couple of weeks he sat down and built a program named Hypermail that would "live" on the same computer that was running a mail server and would generate a threaded copy of an email conversation that could be retrieved from the Web. What emerged was a digital snapshot of the email conversation complete with permanent links that could be bookmarked and archived.

The emergence of the World Wide Web was a life-changing event for Gruber. He was now thirty years old and working at Stanford, and he quickly realized that the Web was a much bigger idea than anything he had worked on previously. He recognized that Tenenbaum was onto something with dramatic potential to change the way people used computers. Tenenbaum had hired a young programmer named Kevin Hughes who had come to the project from Hawaii Community College. Hughes was representative of a new class of programmer that was emerging from a generation who had grown up with computing. He didn't look like he was old enough to drive, but he called himself a "webmaster." Gruber had initially written Hypermail in his favorite programming language, Lisp, and shared it through the software channels that were popular at the time. To Hughes, that approach was dated. He told Gruber that Hypermail had to be rewritten in C and that it had to be given away freely. Gruber convinced Tenenbaum, and then took a weekend to rewrite the program in C. Hughes was right. Once it was freely available on the Web, its use exploded.

It was a major step in bringing the Engelbart-Nelson hypertext vision to life. Overnight, anyone who was running a list server on a Unix computer could drop the program on their computer and their electronic conversations would be instantly available to the broader Internet. It was a powerful lesson for Gruber about how the Internet could be used to leverage a simple idea. EIT was purchased by VeriFone early in 1995, just at the outset of the dot-com era. Two years later VeriFone, known for its point-of-sale terminals, was itself purchased by HP during the run-up of the first Internet bubble, only to be cast out again after the bubble burst. Gruber had left Stanford to join EIT in 1994 but left before EIT was sold the first time to pursue his own ideas. Why stop with email, he wondered? He set out to build on a large chunk of Engelbart's vision and sell it to corporate America.

I n the early 1990s, Engelbart's ideas enjoyed a renaissance at Stanford. The four years Gruber had spent at the university trying to create Feigenbaum's engineering knowledge system from piles of statements of rules and assertions and ontologies hadn't succeeded. In his Hypermail project, Gruber saw a way to build a valuable commercial knowledge system and, in the entrepreneurial heat of the dot-com explosion, set out to create his own company to do so. Berners-Lee had made the original breakthrough when he designed the World Wide Web. It was not just that he had created a working version of the Engelbart-Nelson hypertext system. He had established a system of permanent identifiers for bundles of information that the engineers described as "knowledge objects." That changed everything. It allowed Web developers to create persistent knowledge structures that functioned as usable digital libraries, upon which it was possible to build both artificial intelligence and augmentation systems.

Gruber's idea was to build a "corporate memory," a system that would weave together all the documents that made a modern organization function, making them easy to structure and recall. It was reminiscent of Engelbart's original oN-Line System, but was modernized to take advantage of the power of Berner-Lee's invention. Lotus Notes had been an earlier effort by Ray Ozzie, then a young software designer working on a contract basis for Mitch Kapor at Lotus, but it was stuck in the proprietary world of corporate enterprise software. Now the Internet and new Web standards made it possible to build something with far greater scope.

With another AI researcher, Peter Friedland, and former DARPA program manager Craig Wier, Gruber founded Intraspect in 1996 in Los Altos and became the chief technology officer. In the beginning he worked with one programmer, who had a day job at Stanford. The programmer arrived in the evening after Gruber had worked on the prototype during the day and took over and continued development into the night. As Gruber was leaving at the end of the day, he discussed what he had done and what needed to be completed. They would iterate—it was an effective way to rapidly build a prototype.

The company eventually raised more than $60 million in venture funding and would have as many as 220 employees and a polished product. They were able to quickly build a base of blue-chip companies as well, including GTE, General Motors, KPMG, Boeing, and Siemens. The PC era had transformed the corporation and companies were being run with electronic mail rather than with printed interoffice memos. This shift made it possible to create an inexpensive system that simply "swallowed" every communication that was CC'ed to a special email address. No webmaster or heavy IT presence was necessary. The Intraspect system ingested corporate communications and documents and made them instantly accessible to anyone in a company with access to a personal computer. Desktop folder icons were still the common metaphor for organized documents, and so the Intraspect engineers built a Windows program based on a folder-oriented interface.

In Gruber's mind this was what the future of AI should be. What had started as an effort to model a human brain would shift in focus and end up as an endeavor to model the interactions of a human group. In a sense, this distinction was at the heart of the cultural divide between the two research communities. The AI community began by trying to model isolated human intelligence while the emerging community of human-computer interaction designers followed in Engelbart's augmentation tradition. He had begun by designing a computer system that enhanced the capabilities of small groups of people who collaborated. Now Gruber had firmly aligned himself with the IA community. At the Stanford Knowledge Systems Laboratory, he had interviewed avionics designers and took their insights to heart. There had been an entire era of industrial design during which designers assumed that people would adapt to the machine. Designers originally believed that the machine was the center of the universe and the people who used the machines were peripheral actors. Aircraft designers had learned the hard way that until they considered the human-machine interaction as a single system, they built control systems that led to aircraft crashes. It simply wasn't possible to account for all accidents by blaming pilot error. Aircraft cockpit design changed, however, when designers realized that the pilot was part of the system. Variables like attention span and cognitive load, which had been pioneered and popularized by psychologists, became an integral part first in avionics and, more recently, computer system design.

Gruber thought hard about these issues while he designed the Intraspect query system. He imagined customers, often corporate salespeople, as aircraft pilots and tried to shape the program to avoid deluging them with information. Intraspect demonstrated the system to a J.P. Morgan executive. Gruber performed a simple search and the executive saw relevant recent communications between key employees at the firm with relevant documents attached. His jaw dropped. He could literally see what his company knew. The Intraspect system used a search engine that was engineered to prioritize both the most recent and most relevant documents, which was something that was not yet widely offered by the first generation of Internet search engines.

At the peak of the dot-com era, Intraspect was doing spectacularly well. They had blue-chip customers and footholds in major industries like financial services and a run rate of $30 to $40 million in revenue. They had even written an S-1 in preparation for going public and had moved into a large new building and a prominent logo that was visible from the 101 freeway. Then everything collapsed. Although Intraspect survived the dot-com crash, the meltdown crippled some of its best customers.

September 11, 2001, followed. Overnight, everything changed. By the following March, CFOs at major companies were simply prohibiting the purchase of any product or service that wasn't from a public company. The wave of fear nailed Intraspect. Gruber had spent six years building the company and at first he refused to believe that it was over. They had such strong customers and such a versatile product that they were convinced there must be a way to survive. But

the company had been riding on its ability to leverage professional service firms like the Big Five accounting companies to
sell its product, and those sales channels dried up during the
crash.

Gruber was forced to lay off 60 percent of his company
to stay afloat. In the end, Intraspect died with a whimper.
Although it had advantages over its competitors, the entire
market for collaborative corporate software collapsed. Portal
companies, document management companies, search companies, and knowledge management companies all merged into
one another. In 2003 Intraspect was sold for a fire sale price to
Vignette and that was the end.

Gruber stayed at Vignette for a couple of months and then
took a year off to recharge and think about what he would
do next. He traveled to Thailand, where he went scuba div-
ing and took pictures. He discovered Burning Man, the annual
weeklong gathering in the Nevada desert that attracted tens of
thousands of the Valley's digerati. When Gruber's sabbatical
year ended he was ready to build a new company.

He knew Reid Hoffman, who had by then started LinkedIn,
the business networking company. Because of his experience
at Intraspect, Gruber had good insights into "social software."
The two men had a long series of conversations about Gruber
joining the start-up, which was on track to become one of Silicon Valley's early social networking success stories. Gruber
wanted to focus on design issues and Hoffman was looking for
a new CTO, but in the end the LinkedIn board vetoed the idea
because the company was on the verge of a large investment
round.

Gruber's year of traveling had left him thinking about the
intersection of travel and "collective intelligence" that was coming to life with emergence of "Web 2.0." The Internet had not
only made it possible to create corporate memories, but now

crowdsourcing had become trivial for any human endeavor. Google, of course, was the most spectacular example. The company's PageRank search algorithm exploited human preferences to rank Internet search query results. Through Reid Hoffman, Gruber found a start-up that was planning to compete with TripAdvisor, which at that point was only offering travelers' reviews of hotels. He convinced them that he could bring them a big audience—they just needed to handle the business development side of the project. And so Gruber started over as the vice president of design at this new start-up, although this time he had a team of three engineers instead of sixty. Having a small army of programmers, however, was no longer critical to the success of a company—the Internet had changed everything. Even the smallest start-ups could leverage vastly more powerful development toolkits.

The start-up planned to collect the best trip descriptions that global travelers had to offer. It took them a year to build the service and they unveiled realtravel.com at the 2006 O'Reilly Web 2.0 Conference—an Internet event that had rapidly become the conference of choice for the next wave of so-called social start-ups. Realtravel.com grew fast—it even boasted a couple million unique visitors at one point—but it didn't grow quickly enough, and the company was sold in 2008, just two years after receiving its seed funding. Gruber had left the company before it was sold, having feuded with the CEO—who was color-blind—over issues like what were the best colors on the site's Web pages.

He took another year off. He had worked in various positions at realtravel.com—from writing code to overseeing design—and he needed the time away. When he returned, he used his Silicon Valley network of contacts to look for interesting projects. He was a member of an informal group called the CTO Club, which met regularly, and someone there mentioned a new project at SRI.

The research center had been showered with funding by

DARPA under Tony Tether, who had taken a deep interest in building a software personal assistant. For five years, between 2003 and 2008, the Pentagon agency invested heavily in the idea of a "cognitive assistant." The project would ultimately bring together more than three hundred researchers at twenty-five universities and corporate research laboratories, with SRI playing the role of the integrator for the project. The cognitive assistant, CALO, was in DARPA's tradition of funding blue-sky research that had already created entire industries in Silicon Valley. Workstations, networking, and personal computing had all started as DARPA research projects.

The term "CALO" was inspired by *calonis,* a Latin word meaning "soldier's low servant," or clumsy drudge, but the project also had a significant overlap with Engelbart's original work that was funded by DARPA in the sixties and seventies. CALO was intended to help an office worker with project management: it would organize workers' email, calendars, documents, communication, schedules, and task management. Eventually, there were a number of commercial spin-offs from the CALO project—a smart calendar, a personalized travel guide, and a game development and education company—but they all paled compared to the success of Siri.

Long before the Maker Movement—the Silicon Valley sub-culture extolling an inclusive do-it-yourself approach to technology—gained steam, Gruber's Siri cofounder Adam Cheyer was drafted into that world by his mother. As a child in a Boston suburb, he was restricted to just an hour of television each week, which offered him a brief glimpse of technology that whet his appetite for the latest toys. When he asked his mother to buy toys for him, however, she responded by giving him a stack of cardboard inserts that cleaners used to stiffen shirts. He resorted to tape, glue, and scissors to re-create the toys that he had seen on television, like robots and Rube Goldberg contraptions. It taught Cheyer that with a small dose of imagination, he could make anything he wanted.[2]

As a child he dreamed of becoming a magician. He had read books about the great magicians and thought of them as inventors and tinkerers who tricked others by using technology. Before he was ten he was saving his money to buy books and tricks from the local magic store. Later, he realized that his interest in artificial intelligence was rooted in his love of magic. His favorite eighteenth-century magicians and clockmakers led by Jacques de Vaucanson had built early automata: chess-playing and speaking machines and other mechanical humanoid robots that attempted to illuminate the inner workings of what he, like Gruber, would come to see as the most magical device of all—the human brain.[3]

Although Cheyer knew nothing of Engelbart's legendary NLS, in 1987 he built his own system called HyperDoc while working as an artificial intelligence researcher with Bull, the aerospace firm, in France. He integrated a documentation system into the editor the programmers were using to design their expert systems. That update made it possible to simply click on any function or command to view a related online manual. Having easy access to the software documentation made it simpler for developers to program the computers and reduce the number of bugs. At the time, however, he was unfamiliar with the history of Doug Engelbart's Augmentation Research Center in Menlo Park during the 1960s and 1970s. He had moved to California to get a master's degree in computer science, with a plan to move back to France after graduation. It had been a fun sojourn in California, but the French computer firm would pay for his schooling only if he returned to Europe.

Not long before he was scheduled to return, however, he stumbled across a small blurb advertising a job in an artificial intelligence research laboratory at SRI. The job sounded intriguing and he decided to apply. Before flying to the Bay

Area for the interview, he read extensively on the work of all of the researchers in the group. Between interviews he went into the bathroom to scan his notes in preparation for each appointment. When he arrived, he knew everything that everyone had worked on, who they worked with, and what their views were on different issues. His research paid off. He was hired in the SRI Artificial Intelligence Center.

In the early 1990s, despite the AI Winter, SRI remained a thriving hub for commercial, military, and academic artificial intelligence research, and decades after Shakey, robots were still roaming the halls. When Cheyer arrived at the laboratory, he received a small research grant from a Korean telecom lab run by the South Korean government. The project funding was for a pen and voice control system for the office environment. "Build us one of those," they instructed him.

He decided to build a system that would make it easy to plug in additional capabilities in the future. The system was named Open Agent Architecture, or OAA. It was designed to facilitate what Cheyer thought of as "delegated computing." For example, if a computer needed to answer a question like, "What's Bob's email address?" there was a range of ways that it could hunt for the answer. Cheyer created a language that would make it possible for a virtual software assistant to interpret the task and hunt for the answer efficiently.

In designing his framework, he found that he was at the heart of a debate that was raging between artificial intelligence researchers and the rival human-computer interaction community. One group believed the user needed to be in complete control of the computer and the other group envisioned software agents that could "live" in computer networks and operate on behalf of human users. From the beginning Cheyer had a nuanced view of the ideal human-machine relationship. He thought that humans sometimes like to control systems directly, but often they just want the system to do something on their behalf without bothering them with the details. To that end, his language made it possible to separate what the user wanted the system to do or find from how the task would be accomplished.

Within a year of arriving at SRI, Cheyer was focused on the challenge of actually building a working version of the Knowledge Navigator software avatar that John Sculley had extolled in a futuristic video in 1987. Like Alan Kay, who started out by building "interim" Dynabooks, during the next two decades Cheyer repeatedly developed prototypes, each of which more closely approximated the capabilities of the Knowledge Navigator. He was building software virtual robots, software assistants that were intended to act as much as partners as slaves.

By the end of 1993 he had designed a tablet PC resembling an iPad. No one had developed a touch interface yet and so Cheyer had integrated pen input into his tablet, which allowed it to recognize both handwriting and user gestures, like drawing circles around certain objects to select them. It also had the ability to recognize speech, largely because Cheyer had become adept at the technology equivalent of borrowing a cup of sugar from his neighbors down the hall. He had persuaded the researchers at SRI's Speech Technology and Research Laboratory to install a software connector—known as an API—for his tablet. That allowed him to plug the mainframe-based speech recognition system into his system. SRI's speech technology—which was a research activity that had started with Shakey—would be spun out the next year as a separate start-up, Nuance Communications, which initially pioneered voice applications for call centers. He did the same with SRI handwriting recognition technologies. He built a demonstration system that used voice and pen input to approximate a software secretary. It automated calendar tasks and handled email, contact lists, and databases, and he started experimenting with virtual assistance tasks, like using maps to find restaurants and movie theaters.

Cheyer walked the halls and sampled the different projects

at the laboratory, like natural language understanding, speech recognition, cooperating robots, and machine vision. SRI was his playground and he used it to mash together a remarkably disparate and rich set of computing systems and services—and he did it all before he saw his first Web browser. The World Wide Web was just beginning to filter out into the world. When the NCSA Mosaic browser, the first popular browser that brought the Web to the general public, finally did arrive, it felt like déjà vu.

Cheyer wanted to create an assistant that could provide a computer user with the kind of help he or she might expect to get from an attentive secretary. Although he had started on his own, over the next six years he worked with a small team of programmers and designers and created more than four dozen applications, ranging from intelligent refrigerators that would find recipes and restock themselves to televisions that let you control your home, collaborative robots, and intelligent offices. Ultimately the team would have a significant impact on mobile computing. Fifteen years later, two members of his early research group were key technology executives overseeing the design of the Samsung Galaxy smartphone and three had gone on to Apple to deliver Siri.

Cheyer quietly earned a reputation inside SRI as the "next Engelbart." Eventually he became so passionate about Engelbart's ideas that he kept a photo of the legendary computer scientist on his desk to remind him of his principles. By the end of the 1990s Cheyer was ready for a new challenge. The dot-com era was in full swing and he decided to commercialize his ideas. The business-to-business Internet was exploding and everywhere there were services that needed to be interconnected. His research was a perfect fit for the newly popular idea of loosely coupled control. In a world of networked computers, software that allowed them to cooperate was just beginning to be designed. He was following a similar path to Marty Tenenbaum's, the AI researcher who had created CommerceNet, the company for which Tom Gruber built ontologies.

One of a small group of Silicon Valley researchers who realized early on that the Internet would become the glue that connected all commerce, Cheyer went to a competitor called VerticalNet, where he created a research lab and was soon made VP of engineering. Like Gruber, he was caught up in the dot-com maelstrom. At one point VerticalNet's market value soared to $12 billion on revenues of a little more than $112 million. Of course it couldn't last, and it didn't. He stayed for four years and then found his way back to SRI.

DARPA knocked on Cheyer's door with an offer to head up Tony Tether's ambitious national CALO effort, which DARPA anticipated would draw on the efforts of AI researchers around the country. Usually DARPA would simultaneously fund many research labs and not integrate the results. The new DARPA program, however, called for SRI to marshal all the research into the development of CALO. Everyone would report to the SRI team and develop a single integrated system. Cheyer helped write the initial DARPA proposal, and when SRI received the award, he became engineering architect for the project. CALO was rooted firmly in the traditional world of first-generation symbolic artificial intelligence—planning and reasoning and ontologies—but there was also a new focus on what has been described as "learning in the wild."

CALO had the trappings of a small Manhattan Project. Over four hundred people were involved at the peak, and the project would generate more than six hundred research papers. DARPA spent almost a quarter billion dollars on the effort, making it one of the most expensive artificial intelligence projects in history. Researchers on the CALO project tried to build a software assistant that would possess humanlike adaptability, learn from the person it worked with, and change its behavior accordingly.

When CALO passed its annual system tests, DARPA was enthusiastic. Tether awarded the project an excellence prize, and some of the technology made the transition into navy projects. But Adam Cheyer, as engineering architect, had experienced more than his share of frustrations. John McCarthy had famously asserted that building a "thinking machine" would require "1.8 Einsteins and one-tenth the resources of the Manhattan Project." To put his estimate in perspective, since the Manhattan Project would cost more than $25 billion in current dollars, McCarthy's estimate would mean that CALO was funded with less than one-tenth of what would be needed to build a thinking machine.

For Cheyer, however, the principal obstacle in designing CALO was not lack of funding. Rather it was that DARPA tried to micromanage his progress. Often unable to pursue its own agenda, the rest of the management team would shunt aside Cheyer's ideas. He had a difficult time shepherding the huge number of teams, each of which had its own priorities and received only a small amount of funding from the CALO project. Cheyer's entreaties to work together on a common project that integrated a huge swath of ideas into a new "cognitive" architecture largely fell on deaf ears. The teams listened politely because they were interested in the next round of money, and they would deliver software, but they all wanted to pursue their own projects. In the end there was no way that a large and bureaucratic program could have a direct impact in the real world.

To cope with his frustrations he laid out a series of side projects to work on in 2007. They ranged from efforts to commercialize the CALO technology to the creation, with several friends, of an activists' social network called change.org. It would be a remarkably productive year for Cheyer. With a graduate student, Didier Guzzoni, he used CALO technologies to build a new software development system that eventually became the foundation for Siri. He also put together a small development team that started commercializing various other components of Siri for applications like smartphone calendars and online news reading. He also quietly helped to cofound Genetic Finance, a stealth machine-learning company that built a cluster of more than one million computers to solve financial problems such as predicting the stock market.

In the midst of all of this, Cheyer approached SRI management to ask for some IR & D funding and told them, "I want a little side project where I'm going to build my own CALO the way it should be done." He wanted to build a single integrated system, not a patchwork quilt from dozens of different organizations. SRI agreed, and he named his project "Active Ontologies." He ran it quietly alongside the much larger operation.

The project gained more traction when a key group of SRI technical leaders met for a daylong retreat in Half Moon Bay, a beach town close to the Menlo Park laboratory. There had been growing pressure to commercialize SRI research from the CEO, Curt Carlson, and CALO was an obvious candidate. The retreat was crucial for hashing out answers to basic questions about the goals for the software, like: What should a personal assistant "feel" like? Should they use an avatar design? Avatars had always been a controversial aspect of the design of virtual assistants. Apple's Knowledge Navigator video had envisioned a prim young male with a bow tie who looked a bit like Steve Jobs. The CALO project, on the other hand, did not have an avatar. The developers went back and forth on whether the system should be a chatbot, the kind of personal companion that researchers had explored for decades in programs like Eliza that engaged human users in keyboard "conversations." In the end, they came to a compromise. They decided that nobody was going to sit and chat with a virtual robot all day. Instead, they were going to design a system for people who needed help managing their busy day-to-day lives.

The group came up with the notion of "delight nuggets." Because they were trying to create a humanlike persona, they decided to sprinkle endearing phrases into the software. For example, if a user asked the system about the forecast for the day, the system would answer—and if the forecast indicated that it would rain, the system would add: "And don't forget your umbrella!" The developers wanted to give the user what he or she wanted and to make the design goal about helping them manage their lives—and then to surprise them, just a little bit. Including these phrases added a touch of humanity to the interaction, even though systems did not yet feature speech synthesis and speech recognition.

The 2007 meeting served as a launchpad. SRI's commercialization board gave the team the approval to begin looking for outside money in August. The name Siri ended up working on a wonderful range of levels. Not only did it mean "secret" in Swahili, but Cheyer had once worked on a project called Iris, which was Siri spelled backward. And of course, everyone liked that the name was also a riff on SRI.

In 1987 Apple's chief executive John Sculley gave a keynote address at Educom, the national educational technology conference. He showed a promotional video that a small Apple team had produced to illustrate the idea of something he described as the Knowledge Navigator. At the time, the video, which caught the public's eye (went "viral" in today's parlance), seemed impossibly far out. The Knowledge Navigator was a tour de force that pointed the way to a computing world beyond the desktop computer of the mid-1980s. Knowledge Navigator ultimately spawned a seemingly endless stream of high-tech Silicon Valley "vision statements," including one from Microsoft in 1991 presented by Bill Gates called "Information at Your Fingertips." Yet at that time, the Knowledge Navigator was early to offer a compelling vision for a future beyond desktop personal computing. The video centered on a conversation between an absentminded professor and a perky, bow-tied on-screen avatar as a guide for both the professor's research and his day-to-day affairs. It sketched a future in which computer interaction was no longer based on a keyboard and mouse. Instead, the Knowledge Navigator envisioned a natural conversation with an intelligent machine that both recognized and synthesized human speech.

Brought to Apple as chief executive during the personal computing boom, Sculley started his tenure in 1983 with a well-chronicled romance with Apple's cofounder Steve Jobs. Later, when the company's growth stalled in the face of competition from IBM and others, Sculley fought Jobs for control of the company, and won.

However, in 1986, Jobs launched a new computer company, NeXT. Jobs wanted to make beautiful workstations for college students and faculty researchers. That placed pressure on Sculley to demonstrate that Apple could still innovate without its original visionary. Sculley turned to Alan Kay, who had left Xerox PARC first to create Atari Labs and then came to Apple, for guidance on the future of the computer market. Kay's conversations with Apple's chief executive were summarized in a final chapter in Sculley's autobiographical *Odyssey*. Kay's idea centered on "a wonderful fantasy machine called the Knowledge Navigator,"[4] which wove together a number of his original Dynabook ideas with concepts that would ultimately take shape in the form of the World Wide Web.

Alan Kay would later say that John Sculley had asked him to come up with a "modern Dynabook," which he found humorous, since at the time his original Dynabook still didn't exist. He said that in response to Sculley's request, he had pulled together a variety of ideas from his original Dynabook research and the artificial intelligence community, as well as from MIT Media Laboratory director Nicholas Negroponte, an advocate of speech interfaces.[5] Negroponte had created the Architecture Machine Group at MIT in 1967, in part inspired by the ideas of Ivan Sutherland, whose "Sketchpad" Ph.D. thesis was a seminal work in both computer graphics and interface design.

Historians have underestimated Negroponte's influence on

Apple and the computer industry as a whole. Although Negroponte's "Architecture Machine" idea never gained popular traction, it did have a very specific impact on Bill Atkinson, one of the principal designers of Apple's Lisa and Macintosh computers. Many of the ideas for Lisa and Macintosh were generated from Negroponte's early efforts to envision what the field of architecture would be like with the aid of computers. Negroponte's group created something called "DataLand," a prototype of a visual data management system. In many ways, DataLand was a much broader exploration of how human computer users might interact with information more fluidly. It was certainly broader in scope than the projects at PARC, which focused more narrowly on a creating a virtual desktop. Indeed, Negroponte's goal was expansive. DataLand allowed users to view, in a special room, an immersive information environment back-projected on a giant piece of polished glass as a series of thumbnails representing everything from documents to maps. It was like using a Macintosh or a Windows computer, but having the control screen surround you rather than appear on a small display. It was possible to zoom in on and "fly" through the virtual environment by using a joystick, and when you got close to objects like files, they would talk to you (e.g., "This is Nicholas's Calendar") in a soothing voice. Atkinson visited Negroponte's lab and thought this kind of interface could solve Apple's document filing problem. He wanted to organize documents spatially and place them in proximity to other related documents. Although it was a fascinating concept, it proved unwieldy in practice, and the group returned to something closer to the PARC desktop ideas.

Kay "channeled" ideas that he had gathered in his discussions with Negroponte, passing them on both to Sculley and to the group that created the Knowledge Navigator video. Kay credited Negroponte with playing what he called the "Wayne Gretzky Game"—skating to where the puck was going, rather than where it was. Kay had eagerly read Gordon Moore's early *Electronics* article, which was bold enough to sketch the progress of computing power ten years into the future—1975.[6] He drew the line out to 1995 and beyond. This future-oriented approach meant that he could assume that 3-D graphics would be commercially available within just several decades.

Negroponte represents the "missing link" between Norbert Wiener's early insights into computing and its consequences, the early world of artificial intelligence, and the explosive rise of the personal computer industry during the 1980s. In the late sixties, Negroponte was teaching a course on computer-aided design for architects at MIT. He was not a great fan of lecturing and so had perfected a Tom Sawyer approach to his course—he brought in many guest lecturers. He attracted a dazzling and diverse array of talent. Isaac Asimov, for example, was living in Cambridge at the time and came to Negroponte's class to speak each year, as did Gordon Pask, a British cyberneticist who was traveling widely in U.S. computer research circles in the 1960s and 1970s. If Kay was influenced by Negroponte, he in turn would point to the influence and inspiration of Gordon Pask. At the beginning of the interactive computing era Pask had a broad but generally unchronicled influence on computer and cognitive science research in the United States. Ted Nelson met him in the hallways of the University of Illinois Chicago Circle campus and fell under his spell as well. He described Pask affectionately in his *Computer Lib* manifesto as the "maddest of mad scientists."

In 1968, Negroponte, like many in the computing world, was deeply influenced by Ivan Sutherland's 1963 Ph.D. project, Sketchpad, a graphical and interactive computing tool that pioneered human-computer interaction design. Following in Sutherland's footsteps, Negroponte began work on an "Architecture Machine" that was intended to help human architects build systems that exceeded their individual intellectual grasp. His first effort to build the machine was a software program called URBAN5. The year after he created it, he took a video of his early Architecture Machine project to the London art exhibition known as *Cybernetic Serendipity,* which was held at the Institute of Contemporary Arts. The exhibition had assembled a wild variety of mechanical and computerized art exhibits, including large mobiles created by Gordon Pask, designed with interactive parts to permit viewers to enter into a "conversation" with his installations.

The two met at the exhibition and became good friends. Pask would come to visit the Architecture Machine Group three or four times a year for a week at a time and always stayed as a guest at Marvin Minsky's home. He was a striking character who dressed the part of an Edwardian dandy, complete with a cape, and who occasionally lapsed into double-talk and wordplay. He was squarely in the Norbert Wiener cybernetics tradition, which had taken hold with more force in Europe than in the United States. Pask was also subtly but significantly at odds with the prevailing artificial intelligence world. If AI was about making smart machines that mimicked human capabilities, cybernetics was focused instead on the idea of creating systems to achieve goals.[7] Gordon Pask's insight into the nature of intelligence, which he situated not in the individual but in a conversation between people, strongly influenced Negroponte. Indeed, it was Pask who laid the groundwork for viewing human-machine interactions as conversations that would be later demonstrated by Knowledge Navigator and still later realized in Siri as a conversation: he "conceived human-machine interaction as a form of conversation, a dynamical process, in which the participants learn about each other."[8]

Negroponte was early to grasp Pask's ideas about computer interaction. Later in the 1970s, Pask's ideas also influenced Negroponte's thinking and design at the MIT Media Laboratory as he broadened the original mission of the Architecture Machine Group. Ideas from the Media Lab filtered into Apple's strategic direction because Kay was close to Negroponte and was spending time teaching there. Few noticed it at the time, but Apple's release of Siri as a critical addition to the iPhone

4S in October of 2011 fell within two weeks of Kay's predicted release date for the Knowledge Navigator. The idea traced a direct path from Pask to Negroponte to Kay to the Siri team. A parallel thread ran through Gruber's original work on computing tools for the disabled, to his work on Intraspect, and to the new project at SRI. In the space of just a generation, a wave of computer-mediated communication technology had inaugurated a new way of facilitating collaboration between humans and machines. Gruber recognized that humans had evolved from using tribal communication to written language, and then quickly to using the telephone and computer communications.

Computing had become a prosthesis, not in a bad sense, but rather as a way to augment human capabilities as first foreseen by Vannevar Bush, Licklider, and Engelbart. Intraspect and Hypermail had been efforts to build a cognitive prosthesis for work that needed to go beyond the size of a small tribe. The nature of collaboration was changing overnight. People could have conversations when they weren't in the same room, or even in the same time zone. Simple online email lists like www-talk were being used to develop new Web standards. A permanent archive made it possible for new participants to quickly get up to speed on various issues by reading a record of past conversation.

The idea of the archive would become the guiding principle in the development of Siri. The SRI engineers developed an external memory that provided notes, reminders, schedules, and information, all in the form of a human conversation. The Siri designers adapted the work done on CALO and polished it. They wanted a computer that would take over the task of secretary. They wanted it to be possible to say, "Remind me to call Alan at three thirty or on my drive home."

Just before Cheyer's project was renamed Siri, Gruber would arrive to work with the tiny team at SRI that included Cheyer and Dag Kittlaus. Kittlaus had been managing mobile communications projects at Motorola before coming to SRI. They code named the project HAL, with only a hint of irony.

Cheyer was charming, but he was also fundamentally a highly technical engineer, and for that reason could never be the head of a company. Kittlaus was the opposite. A good-looking, tanned Norwegian executive who straddled the line between technology development and business, he was a quintessentially polished business development operator. He had done early work on the mobile Internet in Europe. Kittlaus arrived with a broad charter, having been asked by the lab's managers to come in as an "entrepreneur-in-residence." There wasn't any particular assignment; he was just supposed to look around and find something promising. It was Kittlaus who found Cheyer. He immediately realized that Cheyer was a hidden gem.

They had first met briefly when Cheyer had been demonstrating prototypes for the wireless industry based on his OAA work in the 1990s. There had been some interest from the telecommunication industry, but Cheyer had realized that there was no way that his toy demos, written in the Prolog artificial intelligence language, would be something that could be used by millions of mobile phone users.

Although SRI later took pains to draw the links between CALO and Siri in order to garner a share of the credit, it was Cheyer who had dedicated his entire career to pursuing the development of a virtual assistant and natural language understanding. When Kittlaus first saw Cheyer's work on Siri in 2007, he told him, "I can make a company out of this!" Cheyer, however, wasn't immediately convinced. He didn't see how Kittlaus could commercialize Siri, but he agreed to help him with the demos. Kittlaus won him over after buying him an iPhone, which had just been released. Cheyer had a very old Nokia and no interest in the new smartphone gadgets. "Play with this!" Kittlaus told him. "This thing is a game changer. Two years from now there will be a competitive response and every handset manufacturer and telco will be desperate to compete with Apple." Since bandwidth would still be slow and screens would still be small, the companies that tried to compete with Apple would have to look for any competitive advantage they could find.

They were planning a start-up and so they began looking for a technical cofounder, but they also needed an outsider to assess the technology. That search led them to Tom Gruber. Cheyer and Kittlaus prepared a simple demo that appeared in Mosaic, the first Web browser, for Gruber. Users could type a question into a search box and it would respond. At the outset he was skeptical.

"I've seen this before, you guys are trying to boil the ocean," he told Cheyer.

The program seemed like a search engine, but then Cheyer began to reveal all the AI components they had integrated into the machine.

Gruber paused. "Wait a moment," he said. "This isn't going to be just a search engine, is it?"

"Oh no," Cheyer responded. "It's an assistant."

"But all you're showing me is a search engine. I haven't seen anything about an assistant," Gruber replied. "Just because it talks to me doesn't mean anything."

He kept asking questions and Cheyer kept showing him hidden features in the system. As he continued the demonstration, Gruber started to run out of steam and fell silent. Kittlaus chimed in: "We're going to put it on phones."

That took Gruber by surprise. At that point, the iPhone had not yet become a huge commercial success.

"This phone is going to be everywhere," Kittlaus said. "This is going to completely change the world. They are going to leave the BlackBerry behind and we want to be on this phone." Gruber had spent his time designing for personal computers and the World Wide Web, not mobile phones, so hearing Kittlaus describe the future of computing was a revelation.

In the mid-2000s, keyboards on mobile phones were a limiting factor and so it made more sense to include speech recognition. SRI had been at the forefront of speech recognition research for decades. Nuance, the largest independent speech recognition firm, got its start as an SRI spin-off, so Cheyer understood the capabilities of speech recognition well.

"It's not quite ready yet," he said. "But it will be."

Gruber was thrilled. Cheyer had been the chief architect of the CALO project at SRI, and Kittlaus had deep knowledge of the mobile phone industry. Moreover, Cheyer had access to a team of great programmers who were equipped with the necessary skills to build an assistant. Gruber realized immediately that this project would reach an audience far larger than anything he had worked on before. In order to succeed, though, the team needed to figure out how to design the service to interact well with humans. From his time at Intraspect and Real Travel, Gruber understood how to build computing systems for use by nontechnical consumers. "You need a VP of design," he told them. It was clear to Gruber that he had the opportunity to work with two of the world's leading experts in their fields, but he had just left an unsuccessful start-up himself. Did he want to sign up again for the crazy world of a start-up so soon?

Why not?

"Do you need a cofounder?" Gruber asked the two men at the end of the meeting.

The core of the team that would build Siri was now in place.

Independently, the three Siri founders had already spent a lot of time pitching investors in the area for funding for earlier projects. In the past, this had been an onerous chore for Gruber, since it required countless visits to venture capitalists who were often uninterested, arrogant, or both. This time their connection to SRI opened the doors to the Valley's blue-chip venture firms. Dag Kittlaus was a master showman, and on their tour of the venture capital firms on Sand Hill Road, he developed a witty and charming pitch. He would take Cheyer and

Gruber in tow to each fund-raising meeting. The men would then be escorted into a conference room and after they introduced themselves, Kittlaus innocently asked the VCs, "Hey, do any of you have one of those newfangled smartphones?" The VCs thrust their hands in their pockets and almost always retrieved Apple's then-brand-new iPhone.

"Do you have the latest apps downloaded?" Kittlaus asked.

Yes.

"Do you have Google search?"

Of course!

Kittlaus then placed a twenty-dollar bill on the table and told the VCs, "If you can answer three questions in five minutes, you can walk away with my money." He then asked the VCs three questions, the answers to which were difficult to search on Google or other similar apps. The venture capitalists listened to the questions and then either said, "Oh, I don't have that app," or made their way through multiple browser pages, following various hyperlinks in an effort to synthesize an answer. Inevitably, the VCs failed to answer even one of the questions in the time allowed, and Kittlaus never lost his money.

It was a clever way for the team to force the potential investors to visualize the need for the missing Siri application. To help them, the team put together fake magazine covers. One of them read: "The End of Search—Enter the Age of the Virtual Personal Assistant." Another one featured an image of Siri crowding Google off the magazine cover. The Siri team also built slides to explain that the Google search was not the end point in the world of information retrieval.

Ultimately the team would be vindicated. Google was slow to come to a broader, more conversational approach to gathering and communicating information. Eventually, however, the search giant would come around to a similar approach. In May of 2013, Amit Singhal, head of the company's "Knowledge" group, which includes search technology, kicked off a product

introduction by proclaiming "the end of search as we know it."
Four years after Siri had arrived, Google acknowledged that
the future of search was conversation. Cheyer's jaw hit the floor
when he heard the presentation. Even Google, a company that
was all about data, had moved away from static search and in
the direction of assistance.

Until they toured Sand Hill for venture capital, Adam Cheyer
had been skeptical that the venture community would buy into
their business case. He kept waiting for VCs to toss them out of
their meetings, but it never happened. At this point, other companies had released less-impressive voice control systems that
had gone bust. General Magic, the once high-flying handheld
computing Apple spin-off, for example, had tried its hand as a
speech-based personal assistant before going out of business
in 2002. Gradually, however, Cheyer realized that if the team
could develop a really good technical assistant, the venture
capitalists and the money would follow.

The team had started looking for money in late 2007 and
they were funded before the end of that year. They had initially visited Gary Morgenthaler, one of Silicon Valley's elder
statesmen and an influential SRI contact, for advice, but Morgenthaler liked the idea so much that he invited them back to
pitch. In the end, the team picked Morgenthaler and Menlo
Ventures, another well-known venture firm.

Before the dot-com era, companies kept their projects
under wraps until they were ready to announce their developments at grand publicity events, but that changed during the
Silicon Valley buildup to the bubble in the late 1990s. There
was a new spirit of openness among more service-oriented new
companies, which shared information freely and raced to be
first to market. The Siri developers, however, decided to stay
quiet; they even used the domain name stealth-company.com
as a placeholder and a tease. They found office space in San
Jose, far away from the other software start-ups that frequently
settled in San Francisco. Having a base in San Jose also made

it easy to find new talent. At the time, technical workers with families were moving to the south end of the Peninsula, and commuting to downtown San Jose was a breeze compared to the trek to Mountain View or Palo Alto.

To build the company culture, Adam Cheyer went out and bought picture frames and handed them out to all of the company's employees. He asked everyone to choose a hero and then put a framed picture of that person on their desks. Then, he asked them to pick a quote that exemplified why that person was important to them. Cheyer hoped this would serve two purposes: it would be interesting to see who people chose, and it would also reflect something about each employee. Cheyer chose Engelbart and attached an early commitment made by the pioneering SRI researcher: "As much as possible, to boost mankind's collective capability for coping with complex, urgent problems." For Cheyer, the quote perfectly expressed the tension between automating and augmenting the human experience. He had always harbored a tiny feeling of guilt about his work as he moved between what he thought of as "people-based" systems and artificial intelligence–based projects. His entire career had vacillated between the two poles. It was 2007, the year that he also helped his friends start the activist site change.org, which fell squarely within the Engelbart tradition, and he believed that Siri was moving along the same path. Gruber had wanted to choose Engelbart as well, but when Cheyer chose him first he fell back on his musical hero, Frank Zappa.

Despite having his project sold to Tymnet in the early 1970s, Doug Engelbart had been brought back into the fold at SRI when Cheyer had arrived, and Cheyer had come to know the aging computer scientist as a father figure and a guiding light. Working on projects that were inspired by Engelbart's augmentation ideas, he had tried to persuade Engelbart that he was working in his tradition. It had been challenging. By the 1990s, Engelbart, who had mapped it all out beginning in the 1960s, was a forlorn figure who felt the world had ignored him. It didn't matter to Cheyer. He saw the power of Engelbart's original vision clearly and he took it with him when he left SRI to build Siri.

In college, Cheyer had begun visualizing goals clearly and then systematically working to achieve them. One day just as they were getting started he wandered into an Apple Store and saw posters with an array of colorfully crafted icons representing the most popular iPhone applications. All of the powerful software companies were there: Google, Pandora, Skype. He focused on the advertising display and said to himself: "Someday Siri is going to have its icon right here on the wall of an Apple Store! I can picture it and I'm going to make this happen."

They went to work. In Gruber's view, the team was a perfect mix. Cheyer was a world-class engineer, Kittlaus was a great showman, and Gruber was someone who could build high-technology demos that wowed audiences. They knew how to position their project for investors and consumers alike. They not only anticipated the kinds of questions people would ask during a demo; they also researched ideas and technology that would have the most crowd appeal. Convincing the observer that the future was just around the corner became an art form unique to Silicon Valley. The danger, of course, was being too convincing. Promising too much was a clear recipe for disaster. Other personal assistants projects had failed, and John Sculley had publicized a grand vision for Knowledge Navigator, which he never delivered. As Siri's developers kicked the project into high gear, Gruber dug out a copy of the Knowledge Navigator video. When Apple had shown it years earlier, it had instigated a heated debate within the user interface design community. Some would argue—and still argue—against the idea of personifying virtual assistants. Critics, such as Ben Shneiderman, insisted that software assistants were both technically and ethically flawed. They argued for keeping human users in direct control rather than handing off decisions to a software valet.

The Siri team did not shy away from the controversy, and it wasn't long before they pulled back the curtain on their project, just a bit. By late spring 2009, Gruber was speaking obliquely about the new technology. During the summer of that year he appeared at a Semantic Web conference and described, point by point, how the futuristic technologies in the Knowledge Navigator were becoming a reality: there were now touch screens that enabled so-called gestural interfaces, there was a global network for information sharing and collaboration, developers were coding programs that interacted with humans, and engineers had started to finesse natural and continuous speech recognition. "This is a big problem that has been worked on for a long time, and we're beginning to see some progress," he told the audience. Gruber also pointed to developments that were on the horizon, like conversational speech between a computer agent and a human and the delegation of tasks to computers—like telling a computer: "Go ahead, make that appointment." Finally, he noted, there was the issue of trust. In the Knowledge Navigator video, the professor had let the computer agent handle calls from his mother. If that wasn't a sign of trust, what was? Gruber hoped his technology would inspire that same level of commitment.

After discussing the technologies forecasted in the Knowledge Navigator video, Gruber teased the audience. "Do we think that this Knowledge Navigator vision is possible today?" he asked. "I'm here to announce"—he paused slightly for effect—"that the answer is still no." The audience howled with laughter and broke into applause. He added, "But we're getting there."

The Siri designers discovered early on that they could quickly improve cloud-based speech recognition. At that point, they weren't using the SRI-inspired Nuance technology, but instead a rival system called Vlingo. Cheyer noticed that when speech recognition systems were placed on the Web, they were exposed to a torrent of data in the form of millions of user queries and corrections. This data set up a powerful feedback loop to train and improve Siri.

The developers continued to believe that their competitive advantage would be that the Siri service represented a fundamental break with the dominant paradigm for finding information on the Web—the information search—exemplified by Google's dramatically successful search engine. Siri was *not* a search engine. It was an intelligent agent in the form of a virtual assistant that was capable of social interaction with humans. Gruber, who was also chief technology officer at Siri, laid out the concepts underlying the service in a series of white papers in the form of technical presentations. Finding information should be a conversation, not a search, he argued. The program should be capable of disambiguating questions to refine the answers to human questions. Siri would provide services—like finding movies and restaurants—not content. It would act as a highly personalized broker for the human user. In early 2010 the Siri team put together an iPhone demonstration for their board of directors. Siri couldn't speak yet, but the program could interpret spoken queries and converse by responding to human queries in natural language sentences that were displayed in cartoonlike bubbles on screen. The board was enthusiastic and gave the developers more time to tune and polish the program.

In February of 2010, the tiny start-up released the program on the iPhone App Store. They received early positive reviews from the Silicon Valley digerati. Robert Scoble, one of the Valley's high-profile technology bloggers, referred to it as "the most useful thing that I've seen so far this year." Faint praise perhaps—it was still very early in the year.

Gruber was away at a technology retreat during the release

and had almost no access to the Web when the product was first available. He had to rely on secondhand reports—"Dude, have you seen what's happening to your app?!"—to keep up.

It got better. Thanks to a clever decision to place the application in a less obvious category on the App Store—Lifestyle— the Siri Assistant immediately shot right to the top of the category. It was one of the tricks Gruber had learned during his time at Real Travel—the art of search engine optimization. Although they had introduced Siri on the iPhone, Kittlaus had negotiated a spectacular agreement with Verizon, which did not yet carry the iPhone. He described it as "the greatest mobile deal in history." The deal guaranteed that Siri would be on every new Verizon phone, which meant that the software would become the poster child for the Android smartphone. The deal was almost set in stone when Kittlaus received a call on his cell phone.

"Hi, Dag," the caller said. "This is Steve Jobs."

Kittlaus was momentarily stunned. "How did you get this phone number?" he asked.

"It's a funny story," Jobs replied. He hadn't had any idea how to find the small development team, but he had hunted around. Because every iPhone developer had to supply a phone number to the App Store, Apple's CEO found Kittlaus's number in his developer database.

The team's first foray into the legendary "reality distortion field"—Jobs's personal brand of hypnotic charisma—wasn't promising. Jobs invited the trio of Siri developers to his house in the heart of old Palo Alto. Jobs's home was a relatively low-key 1930s Tudor-style set next to an empty lot that he had converted into a small grove of fruit trees and a garden. They met in the living room, which was sparsely furnished, like much of Jobs's home, and featured an imposing Ansel Adams original.

Jobs presented the trio with a dilemma. They had all been successful in Silicon Valley, but none of them had yet achieved the career-defining IPO. The Siri team—and certainly their board members—thought it was very possible that they would receive a huge public stock offering for Siri. Jobs made it clear that he wanted to acquire Siri, but at that juncture the team wasn't planning to sell. "Thank you very much," they told him, and then left.

Several weeks later Apple was back. They were once again invited to Jobs's home, where Jobs, then clearly sick despite continuing to publicly deny it, turned on the charm. He promised them an overnight market of one hundred million users—with no marketing and no business model. Or, Jobs said, they could roll the dice, try to be the next Google, and slog it out. The Siri team also understood that if they went with Verizon, they would run the risk of being shut out of the iTunes Store. Steve didn't have to say it, but it was clear that they had to choose which half of the market they wanted.

Jobs's offer sold them, but it didn't immediately sell the board, which was by now eager for an IPO exit. The three founders had to reverse ground and persuade their board members. Ultimately the investors were convinced; Jobs's offer was lucrative enough and offered much lower risk.

Soon after Apple acquired Siri in April of 2010, the Siri team moved into the very heart of the office space for Apple's design group, on half of the top floor of Infinite Loop 2. Although Apple could have licensed Nuance to convert speech directly to text—which Google later did—Jobs decided that Apple would undertake the more ambitious task of placing an intelligent assistant software avatar on the iPhone. Siri helped solve another major problem that Apple had with its new iPhone and iPad. Glass screens and multitouch control could replace a keyboard and mouse for navigation through screens, but they did not work well for data entry. This was a weak point, despite Jobs's magnificent demonstration of text entry and auto-correction during the first product introduction. Speech entry of single words or entire sentences is many times more rapid than painstakingly entering individual words by poking at the screen with a finger.

In the beginning, however, the project was met with resistance within the company. Apple employees would refer to the technology as "voice control," and the Siri team had to patiently explain that their project had a different focus. The Siri project didn't feed into the "eye candy" focus at Apple—the detailed attention of software and hardware design that literally defined Apple as a company—but was instead about providing customers with reliable and invisible software that worked well. But many engineers in the software development organization at Apple thought that if Steve—and later on one of his top lieutenants, Scott Forstall—didn't say "make it happen," they didn't need to work on that project. After all, Apple was not recognized as a company that developed cloud-computing services. Why reinvent the wheel? An assistant or simply voice control? After all, how much difference would it really make? In fact, people were dying while reading email and "driving while intexticated," so presenting drivers with the ability to use their phones safely while driving made a tremendous difference.

When Apple's project management bureaucracy balked at the idea of including the ability to send a hands-free text message in the first version of the software, Gruber, who had taken the role of a free-floating technical contributor after the acquisition, said he would take personal responsibility for completing the project in time for the initial Apple Siri launch. He decided it was a "put your badge on the table" issue. With just a summer intern in tow, he worked on all of the design and prototyping for the text messaging feature. He begged and borrowed software engineers' time to help build the system. In the end, it was accepted. At the time of Siri's launch, it was possible to send and receive texts without touching the iPhone screen.

Not everything went as smoothly, however. The Siri team also wanted to focus on what he called "attention management." The virtual personal assistant should also help people remember their "to-do list" in an "external memory" so they wouldn't have to. The original Siri application had an elaborate design for what the team described as "personal memory": it wove an entire set of tasks together in the right order, prodding the user at each step like a good secretary. In the race to bring Siri to the iPhone, however, much of the deeper power of the service was shelved, at least temporarily. The first iteration of Siri only included a small subset of what the team had originally created.

In his final act in the computing world, Steve Jobs had come down emphatically on the side of the forces of augmentation and partnership. Siri was intended to be a graceful, understated model for the future collaboration between humans and machines, and it marked the beginning of a sea change at Apple that would take years to play out. The project also came together in a furious rush, and sadly Jobs died the day after Siri's debut. The product launch event in October 2011 thus had to acknowledge a muted counterpoint in what was otherwise a glorious crowning moment to their rocket-fast three-year crusade. Naturally, there was a shared feeling of triumph. On the morning of Siri's unveiling, Cheyer found himself back in an Apple Store. He walked up to the store and next to the front door was a giant plasma display that read: "Introducing Siri!"

It was almost midnight in Grand Central Station on a spring night in 1992. An elderly man wearing a blue *New York Times* windbreaker was leaning on a cane on the platform, waiting for a train to Westchester County. I had been at the *Times* for several years and I was puzzled by the ghostly figure. "Do you work for the paper?" I asked.

Many years earlier, he said, he had been a typesetter at the *Times*. In 1973 his union negotiated a deal to phase out workers' jobs while the company implemented computerized printing systems in exchange for guaranteed employment until retirement. Although he had not worked for more than a decade, he still enjoyed coming to the press in Times Square and spending his evenings with the remaining pressmen as they produced the next day's paper.

Today, the printer's fate remains a poignant story about labor in the face of a new wave of AI-based automation technologies. His union first battled with newspaper publishers in the 1960s and struck a historic accommodation in the 1970s. Since then, however, the power of unions has declined dramatically. In the past three decades, the unionized percentage of the U.S. workforce has fallen from 20.1 to 11.3 percent. Collective bargaining will not play a significant role in defending workers' jobs against the next wave of computerization. Printers and typographers in particular were highly skilled workers who fell prey to the technical advances of a generation of minicomputers during the 1970s, and the cost of computing plummeted as technologies shifted from transistor-based machines to lower-cost integrated circuits. Today, the lone typesetter's soft landing is an extraordinary rarity.

There is evidence that the 2008 recession significantly accelerated the displacement of workers by computerized systems. Why rehire workers when your company can buy technology that replaces them for less cost? A 2014 working paper released by the National Bureau of Economic Research confirmed the trend, and yet Henry Siu, an associate professor at the University of British Columbia and one of the authors of the report, clung to the conventional Keynesian view on technological unemployment. He explained: "Over the very long run, technological progress is good for everybody, but over shorter time horizons, it's not that everybody's a winner."[1] It is probably worth noting that Keynes also pointed out that in the long run, we are all dead.

Indeed, Keynes's actuarial logic is impeccable, but his economic logic is now under assault. There is an emerging perspective among technologists and some economists that Keynesian assumptions about technological unemployment— that individual jobs are lost but the overall amount of work stays constant—no longer hold true. AI systems that can move, see, touch, and reason are fundamentally altering the equation for human job creation. The debate today is not whether AI systems will arrive, but when.

It is still possible that history will vindicate the Keynesians. Modern society may be on the cusp of another economic transformation akin to the industrial revolution. It is conceivable that social forces like crowdsourcing and the Internet-enabled reorganization of the workforce will remake the U.S. economy in ways that are now inconceivable. The Internet has already created new job categories like "search engine optimization," and there will certainly be other Internet-enabled and unexpected new job categories in the future.

However, if there is a new employment boom coming, it is still over the horizon. The Bureau of Labor Statistics projections now predict that U.S. job growth will be primarily influenced by the aging of American society, not by technological advances that displace and create jobs. The BLS predicts that of the 15.6 million jobs that will be created by 2022, 2.4 million of those jobs will be in the health-care and elder-care sectors. It is striking that among new types of jobs, those based on technological advances and innovation will account for a relatively small portion of overall job growth according to the BLS, and of those, software developers were highest ranked at twenty-sixth, with just 139,000 new jobs projected by 2022.[2] The BLS projections suggest that technology will not be a fount of economic growth, but will instead pose a risk to all routinized jobs and skill-based jobs that require the ability to perform diverse kinds of "cognitive" labor, from physicians to reporters to stockbrokers.

Still, despite fears of a "jobs apocalypse," there is another way to consider the impact of automation, robotics, and AI on society. Certainly AI and robotics technologies will destroy a vast number of jobs, but they can also be used to extend humanity. Which path is taken will be determined entirely by individual human designers. Tandy Trower is a software engineer who once oversaw armies of software engineers at Microsoft Corporation, but now works from a cramped office in South Seattle. The four-room shop might be any Silicon Valley garage start-up. There are circuit boards and computers strewn in every direction, and there are robots. Many of them are toys, but several look suspiciously like extras from the movie *Robot & Frank*. The idea of developing a robot to act as a human caregiver speaks directly to the tensions between AI and IA approaches to robotics.

How will we care for our elderly? For some, integrating robots into elder care taps into a largely unmined market and offers roboticists the chance to orient their research toward a social good. Many argue that there is a shortage of skilled caregivers and believe that the development of robots that will act as companions and caregivers is a way of using artificial intelligence to ward off one of the greatest hazards of old age—loneliness and isolation.

The counterpoint to this argument is that there is not really a shortage of caregivers but rather a shortage in society's willingness to allocate resources for tasks such as caregiving and education. "Of course we have enough human caregivers for the elderly. The country—and the world—is awash in underemployment and unemployment, and many people find caregiving to be a fulfilling and desirable profession. The only problem is that we—as a society—don't want to pay caregivers well and don't value their labor," writes Zeynep Tufekci, a social scientist at the University of North Carolina at Chapel Hill.[3] Tufekci was responding to an essay written by Louise Aronson, a gerontologist at the University of California, San Francisco who argued that there is an urgent need for robot caregivers to perform tasks that range from watching over the health of elder patients, organizing their lives, and serving as companions. She describes making house calls and staying much longer than she should for each patient as she is forced to play the dual role of caregiver and companion.[4] Tufekci envisions a society in which a vast army of skilled human doctors will be trained to spend time with the elderly. Sadly, as she notes, we live in a world that places more value on the work of stockbrokers and lawyers than nursing aides and teachers. In the end, however, this argument is not about technology. Once, in agrarian communal societies, families cared for their elders. In Western society, that is frequently no longer the case, and it is inconceivable that we will return to any kind of geographically centralized extended family structure soon.

Still, Tufekci's challenge poses several questions.

First, will robots ever approximate the care of a human stranger? There are many horror stories about elder-care treatment in modern nursing homes and care facilities. Tufekci argues that every elder deserves the attention of an educated, skilled, and compassionate Dr. Aronson. However, if that doesn't happen, will increasingly low-cost robots make life for elders better or worse? The vision of an aging population locked away and "watched over by machines of loving grace" is potentially disturbing. Machines may eventually look, act, and feel as if they are human, but they are decidedly not.

However, robots do not need to entirely replace human caregivers in order to help the elderly. For example, there could be a web of interconnected robots that make it possible for elders who are isolated to build a virtual human community on the Internet. Perhaps shut-in elders will be the most loyal users of augmented reality technologies being designed by Magic Leap, Microsoft, and others. The possibility of virtual caregivers is a compelling idea for those who are physically infirm.

Today Tandy Trower places himself squarely in the augmentation camp. He came to robotics as a member of Bill Gates's technical staff at Microsoft. Gates was touring college campuses during 2006 and realized that there was an intense interest in robotics at computer science departments around the country. Everywhere he went, he watched demonstrations of robotics research. After one of his trips, he came back and asked Trower to put together a proposal for a way that Microsoft might become more active in the emerging robotics industry. Trower wrote a sixty-page report calling on Microsoft to create a group that built software tools to develop robots. Microsoft gave Trower a small group of researchers and he went off to build a simulator and a graphical programming language. They named it the Microsoft Robotics Developer Studio.

Then, however, Gates retired to start his foundation, and everything changed at Microsoft. The new chief executive, Steve Ballmer, had a very different focus. He was more concerned about making money and less willing to take risks. Through Microsoft veteran and chief strategy officer Craig Mundie, he sent Trower a clear message: tell me how Microsoft is going to make money on this.

Ballmer was very clear: he wanted a business that generated one billion dollars in revenue annually within seven years. Microsoft had industrial robotics partners, but these partners had no interest in buying software from Microsoft—they already had their own software. Trower started looking for different industries that might be interested in purchasing his software. He looked at the automotive industry, but Microsoft already had an automotive division. He looked at the science education market, but it didn't have much revenue potential. It seemed too early to pitch a telepresent robot. The more he looked, the more he considered the problem of aging and elder care. "Wow!" he thought to himself. "Here is a market that is going to explode in the next twenty or thirty years." Today in the United States more than 8.5 million seniors require some kind of assistance, and that number will increase to more than 21 million in the next two decades.

There was an obvious need for robotic assistance in elder care, and no major players were angling for that market. Despite his enthusiasm, however, Trower wasn't able to persuade Mundie or Ballmer that Microsoft should invest in the idea. Ballmer was interested in shrinking the range of Microsoft investments and focusing on a few core items.

"I have to do this," Trower thought to himself. And so in late 2009, he left Microsoft after twenty-eight years and founded Hoaloha Robotics—the word *hoaloha* translates from Hawaiian as "friend"—with the intent of creating a mobile elder-care robot at a reasonable cost. Half a decade later, Trower has developed a four-foot-tall robotic prototype, affectionately known as Robby. It isn't a replacement for a human caregiver, but it will be able to listen and speak, help with medicine, relay messages, and act as a telepresence when needed. It doesn't walk—it rolls on a simple wheel assembly that allows it to move fluidly in any direction. Instead of arms, it has a tray whose height it can adjust. This allows Robby to perform certain tasks, like picking up dropped items.

Trower does not think that Robby will displace human workers. Rising costs and a shrinking supply of workers will instead create a situation in which a helper robot can extend the capabilities of both human patients and helpers. Human caregivers already cost $70,000 or more a year, Trower argues, and a low-cost robot will actually extend assistance to those who cannot afford it.

Ignoring Tufekci's fears, Trower has focused his engineering skills on extending and assisting humans. But when will these machines meet our expectations for them? And how will those who are cared for greet them? These remain open questions, although there is a wealth of anecdotal evidence that suggests that, as speech recognition and speech synthesis technologies continue to improve, as sensors fall in cost, and as roboticists develop more agile machines, we will gratefully accept them. Moreover, for an Internet-savvy generation that has grown up with tablets, iPhones, and Siri, caregiving machines will seem like second nature. Robots—elder-care workers, service workers, drivers, and soldiers—are an inevitability. It is more difficult, however, to predict our relationship with these robots. Tales such as that of the golem weave the idea of a happy slave that serves our every desire deeply into our psyches as well as our mythology. In the end the emergence of intelligent machines that largely displace human labor will undoubtedly instigate a crisis of human identity.

For now, Trower has focused on a clear and powerful role for robots as assistants for the infirm and the elderly. This is an excellent example of AI used directly in the service of humans, but what happens if AI-based machines spread quickly through the economy? We can only hope that the Keynesians are vindicated—in the long run.

The twin paths of AI and IA place a tremendous amount of power and responsibility in the hands of the two communities of designers described in this book. For example, when Steve Jobs set out to assemble a team of engineers to reinvent personal computing with Lisa and the Macintosh, he had a clear goal in mind. Jobs thought of computing as a "bicycle for our minds." Personal computing, which was initially proposed by a small group of engineers and visionaries in the 1970s, has since then had a tremendous impact on the economy and the modern workforce. It has both empowered individuals and unlocked human creativity on a global scale.

Three decades later, Andy Rubin's robotics project at Google is representative of a similar small group of engineers who are advancing the state-of-the-art of robots. Rubin set out with an equally clear—if dramatically different—vision in mind. When he started acquiring technology and talent for Google's foray into robotics, he described a ten- to fifteen-year-long effort to radically advance an array of developments in robotics, from walking machines to robot arms and sensor technology. He sketched a vision of bipedal Google delivery robots arriving at homes by sitting on the back of Google cars, from which they would hop off to deliver packages.

Designing humans either into or out of computer systems is increasingly possible today. Further advances in both artificial intelligence and augmentation tools will confront roboticists and computer scientists with clear choices about the design of the systems in the workplace and, increasingly, in the surrounding world. We will soon be living—either comfortably or uncomfortably—with autonomous machines.

Brad Templeton, a software designer and consultant to the Google car project, has asserted, "A robot will be truly autonomous when you instruct it to go to work and it decides to go to the beach instead."[5] It is a wonderful turn of phrase, but he has conflated self-awareness with autonomy. Today, machines are beginning to act without *meaningful* human intervention, or at a level of independence that we can consider autonomous. This level of autonomy poses difficult questions for designers of intelligent machines. For the most part, however, engineers ignore the ethical issues posed by the use of computer technologies. Only occasionally does the community of artificial intelligence researchers sense a quiver of foreboding.

At the Humanoids 2013 conference in Atlanta, which focused on the design and application of robots that appear humanlike, Ronald Arkin, a Georgia Tech roboticist, made a passionate plea to audiences in his speech "How to NOT Build a Terminator." He reminded the group that in addition to his famous three laws, Asimov later added the fundamental "zeroth" law of robotics, which states, "A robot may not harm humanity, or, by inaction, allow humanity to come to harm."[6] Speaking to a group of more than two hundred roboticists and AI experts from universities and corporations, Arkin challenged them to think more deeply about the consequences of automation. "We all know that [the DARPA Robotics Challenge] is motivated by urban seek-and-destroy," he said sardonically, adding, "Oh no, I meant urban search-and-rescue."

The line between robots as rescuers and enforcers is already gray, if it exists at all. Arkin showed clips from sci-fi movies, including James Cameron's 1984 *The Terminator*. Each of the clips depicted evil robots performing tasks that DARPA has specified as part of its robotics challenge: clearing debris, opening doors, breaking through walls, climbing ladders and stairs, and riding in utility vehicles. Designers can exploit these capabilities either constructively or destructively, depending on their intent. The audience laughed nervously—but Arkin refused to let them off the hook. "I'm being facetious," he said, "but I'm just trying to tell you that these kinds of technologies you are developing may have uses in places you may not have fully envisioned." In the world of weapons design, the potential for unexpected consequences has long been true for what are described as "dual-use" technologies, like nuclear power, which can be used to produce both electric power and weapons. Now it is also increasingly true of robotics and artificial intelligence technologies. These technologies are dual-use not just as weapons, but also in terms of their potential to either augment or replace humans. Today, we are still "in the loop"— machines that either replace or augment humans are the product of human designers, so the designers cannot easily absolve themselves of the responsibility for the consequences of their inventions. "If you would like to create a Terminator, then I would contend: Keep doing what you are doing, because you are creating component technologies for such a device," Arkin said. "There is a big world out there, and this world is listening to the consequences of what we are creating."

The issues and complications of automation have extended beyond the technical community. In a little-noted, unclassified Pentagon report entitled "The Role of Autonomy in DoD Systems,"[7] the report's authors pointed out the ethical quandaries involved in the automation of battle systems. The military itself is already struggling to negotiate the tension between autonomous systems, like drones, that promise both accuracy and cost efficiency, and the consequences of stepping ever closer to the line where humans are no longer in control of decisions on life and death. Arkin has argued elsewhere that, unlike human soldiers, autonomous war-fighting robots might have the advantage that they wouldn't feel a threat to their personal safety, which could potentially reduce collateral damage and avoid war crimes. This question is part of a debate that dates back at least to the 1970s, when the air force generals who controlled the nation's fleets of strategic bombers used the human-in-the-loop argument—that it was possible to recall a bomber and use human pilots to assess damage—in an attempt to justify the value of bomber aircraft in the face of more modern ballistic missiles.

But Arkin also posed a new set of ethical questions in his talk. What if we have moral robots but the enemy doesn't? There is no easy answer to that question. Indeed, increasingly intelligent and automated weapons technologies have inspired the latest arms race. Adding inexpensive intelligence to weapons systems threatens to change the international balance of power between nations.

When Arkin concluded his talk at the stately Historic Academy of Medicine in Atlanta, Gill Pratt, the DARPA director of the agency's Robotics Challenge, was one of the first to respond. He didn't refute Arkin's point. Instead, he acknowledged that robots are a "dual-use" technology. "It's very easy to pick on robots that are funded by the Defense Department," he said. "It's very easy to pick on a robot that looks like the Terminator, but in fact with dual-use being everywhere, it really doesn't matter. If you're designing a robot for health care, for instance, the autonomy it needs is actually in excess of what you would need for a disaster response robot."[8] Advanced technologies have long posed questions about dual-use. Now, artificial intelligence and machine autonomy have reframed the problem. Until now, dual-use technologies have explicitly required that humans make ethical decisions about their use. The specter of machine autonomy either places human ethical decision-making at a distance or removes it entirely.

In other fields, certain issues have forced scientists and technologists to consider the potential consequences of their work, and many of those scientists acted to protect humanity. In February of 1975, for example, Nobel laureate Paul Berg encouraged the elite of the then new field of biotechnology to meet at the Asilomar Conference Grounds in Pacific Grove, California. At the time, recombinant DNA—inserting new genes into the DNA of living organisms—was a fledgling development. It presented both the promise for dramatic advances in medicine, agriculture, and new materials and the horrifying possibility that scientists could unintentionally bring about the end of humanity by engineering a synthetic plague. For the scientists, the meeting led to an extraordinary resolution. The group recommended that molecular biologists refrain from certain kinds of research and embark on a period of self-regulation during which they would pause their research while the scientists considered how to make it safe. To monitor the field, biotechnologists set up an independent committee at the National Institutes of Health to review research. After a little more than a decade, the NIH had gathered sufficient evidence from a wide array of experiments to suggest that it should lift the restrictions on research. It was a singular example of how society might thoughtfully engage with the consequences of scientific advance.

Following in the footsteps of the biologists, in February of 2009, a group of artificial intelligence researchers and roboticists also met at Asilomar to discuss the progress of AI after decades of failure. Eric Horvitz, the Microsoft AI researcher who was serving as president of the Association for the Advancement of Artificial Intelligence, called the meeting. During the previous five years, the researchers in the field had begun discussing twin alarms. One came from Ray Kurzweil, who had heralded the relatively near-term arrival of computer superintelligences. Bill Joy, a founder of Sun Microsystems, also offered a darker view of artificial intelligence. He wrote a *Wired* magazine article that detailed a trio of technology threats from the fields of robotics, genetic engineering, and nanotechnology.[9] Joy believed that the technologies represented a triple threat to human survival and he did not see an obvious solution.

The artificial intelligence researchers who met at Asilomar chose to act less cautiously than their predecessors in the field of biotechnology. The group of computer science and robotics luminaries, including Sebastian Thrun, Andrew Ng, Manuela Veloso, and Oren Etzioni, who is now the director of Paul Allen's Allen Institute for Artificial Intelligence, generally discounted the possibility of superintelligences that would surpass humans as well as the possibility that artificial intelligence might spring spontaneously from the Internet. They agreed that robots that can kill autonomously have already been developed, yet, when it emerged toward the end of 2009, the group's report proved to be an anticlimax. The field of AI had not yet arrived at the moment of imminent threat. "The 1975 meeting took place amidst a recent moratorium on recombinant DNA research. In stark contrast to that situation, the context for the AAAI panel is a field that has shown relatively graceful, ongoing progress. Indeed, AI scientists openly refer to progress as being somewhat disappointing in its pace, given hopes and expectations over the years,"[10] the authors wrote in a report summarizing the meeting.

Five years later, however, the question of machine autonomy emerged again. In 2013, when Google acquired DeepMind, a British artificial intelligence firm that specialized in machine learning, popular belief held that roboticists were very close to building completely autonomous robots. The tiny start-up had produced a demonstration that showed its software playing video games, in some cases better than human players. Reports of the acquisition were also accompanied by the claim that Google would set up an "ethics panel" because of concerns about potential uses and abuses of the technology. Shane Legg, one of the cofounders of DeepMind, acknowledged that the technology would ultimately have dark consequences for the human race. "Eventually, I think human extinction will probably occur, and technology will likely play a part in this."[11] For an artificial intelligence researcher who had just reaped hundreds of millions of dollars, it was an odd position to take. If someone believes that technology will likely evolve to destroy humankind, what could motivate them to continue developing that same technology?

At the end of 2014, the 2009 AI meeting at Asilomar was reprised when a new group of AI researchers, funded by one of the Skype founders, met in Puerto Rico to again consider how to make their field safe. Despite a new round of alarming statements about AI dangers from luminaries such as Elon Musk and Stephen Hawking, the attendees wrote an open letter that notably fell short of the call to action that had been the result of the original 1975 Asilomar biotechnology meeting.

Given that DeepMind had been acquired by Google, Legg's public philosophizing is particularly significant. Today, Google is the clearest example of the potential consequences of AI and IA. Founded on an algorithm that efficiently collected human knowledge and then returned it to humans as a powerful tool for finding information, Google is now engaged in building a robot empire. The company will potentially create machines that replace human workers, like drivers, delivery personnel, and electronics assembly workers. Whether it will remain an "augmentation" company or become a predominately AI-oriented organization is unclear.

The new concerns about the potential threat from AI and robotics evoke the issues that confronted the fictional Tyrell Corporation in the science-fiction movie *Blade Runner,* which raised the ethical issues posed by the design of intelligent machines. Early in the movie Deckard, a police detective, confronts Rachael, an employee of a firm that makes robots, or replicants, and asks her if an artificial owl is expensive. She suggests that he doesn't believe the company's work is of value. "Replicants are like any other machine," he responds. "They're either a benefit or a hazard. If they're a benefit, it's not my problem."[12]

How long will it be before Google's intelligent machines, based on technologies from DeepMind and Google's robotics division, raise the same questions? Few movies have had the cultural impact of *Blade Runner*. It has been released seven different times, once with a director's cut, and a sequel is on the docket. It tells the story of a retired Los Angeles police detective in 2019 who is recalled to hunt down and kill a group of genetically engineered artificial beings known as replicants. These replicants were originally created to work off-planet and have returned to Earth illegally in an effort to force their designer to extend their artificially limited life spans. A modern-day *Wizard of Oz,* it captured a technologically literate generation's hopes and fears. From the Tin Man, who gains a heart and thus a measure of humanity, to the replicants who are so superior to humanity that Deckard is ordered to terminate them, humanity's relations to robots have become the defining question of the era.

These "intelligent" machines may never be intelligent in a human sense or self-aware. That's beside the point. Machine intelligence is improving quickly and approaching a level where it will increasingly offer the compelling *appearance* of intelligence. When it opened in December 2013, the movie *Her* struck a chord, most likely because millions of people already interact with personal assistants such as Apple's Siri. *Her*-like interactions have become commonplace. Increasingly as computing moves between desktops and laptops and becomes embedded in everyday objects, we will expect them to communicate intelligently. In the years while he was designing Siri and the project was still hidden from the public eye, Tom Gruber referred to this trend as "intelligence at the interface." He felt he had found a way to blend the competing worlds of AI and IA.

And indeed, the emergence of software-based intelligent assistants hints at a convergence between the work in disparate communities of AI and human-computer interaction designers. Alan Kay, who conceived of the first modern personal computer, has said that in his early explorations of computer interfaces, he was working roughly ten to fifteen years in the future, while Nicholas Negroponte, one of the first people to explore the ideas of immersive media, virtual reality, and conversational interfaces, was working twenty-five to thirty years in the future. Like Negroponte, Kay asserts that the best computerized interfaces are the ones that are closer to theater, and the best theater draws the audience into its world so completely that they feel as if they are part of it. That design focus on interactive performance points directly toward interactive systems that will function more as AI-based "colleagues" than computerized tools.

How will these computer avatars transform society? Humans are already spending a significant fraction of their waking hours either interacting with other humans through computers or directly interacting with humanlike machines, either in fantasy and video games or in a plethora of computerized assistance systems that range from so-called FAQbots to Siri. We even use search engines in our everyday conversations with others.

Will these AI avatars be our slaves, our assistants, our colleagues, or some mixture of all three? Or more ominously, will they become our masters? Considering robots and artificial intelligences in terms of social relationships may initially seem implausible. However, given that we tend to anthropomorphize our machines, we will undoubtedly develop social relationships with them as they become increasingly autonomous. Indeed, it is not much different to reflect on human-robot relations than it is to consider traditional human relations with slaves, who have been dehumanized by their masters throughout history. Hegel explored the relationship between master and slave in *The Phenomenology of Spirit* and his ideas about the "master-slave dialectic" have influenced other thinkers ranging from Karl Marx to Martin Buber. At the heart of Hegel's dialectic is the insight that both the master and the slave are dehumanized by their relationship.

Kay has effectively translated Hegel for the modern age.

Today, a wide variety of companies are developing conversational computers like Siri. Kay argues that as a consequence, designers should aim to create programs that function as colleagues rather than servants. If we fail, history hints at a disturbing consequence. Kay worried that building intelligent "assistants" might only recapitulate the problem the Romans faced by letting their Greek slaves do their thinking for them. Before long, those in power were unable to think independently.

Perhaps we have already begun to slip down a similar path. For example, there is growing evidence that reliance on GPS for directions and for correction of navigational errors hinders our ability to remember and reason spatially, which are more generally useful survival skills.[13] "When people ask me, 'are computers going to take over the world?'" Kay said, "For most people they already have, because they have ceded authority to them in so many different ways."

That hints at a second great problem: the risk of ceding individual control over everyday decisions to a cluster of ever more sophisticated algorithms. Not long ago, Randy Komisar, a veteran Silicon Valley venture capitalist, sat in a meeting listening to someone describe a Google service called Google Now, the company's Siri competitor. "What I realized was that people are dying to have an intelligence tell them what they should be doing," he said. "What food they should be eating, what people they should be meeting, what parties they should be going to." For today's younger generation, the world has been turned upside down, he concluded. Rather than using computers to free them up to think big thoughts, develop close relationships, and exercise their individuality and creativity and freedom, young people were suddenly so starved for direction that they were willing to give up that responsibility to an artificial intelligence in the cloud. What started out as Internet technologies that made it possible for individuals to share preferences efficiently has rapidly transformed into a growing array of algorithms that increasingly dictate those preferences for them. Now the Internet seamlessly serves up life directions. They might be little things like finding the best place nearby for Korean barbecue based on the Internet's increasingly complete understanding of your individual wants and needs, or big things like an Internet service arranging your marriage—not just the food, gifts, and flowers, but your partner, too.

The tension inherent in AI and IA perspectives was a puzzle to me when I first realized that Engelbart and McCarthy had set out to invent computer technologies with radically different goals in mind. Obviously they represent both a dichotomy and a paradox. For if you augment a human with computing technology, you inevitably displace humans as well. At the same time, choosing one side or another in the debate is an ethical choice, even if the choice isn't black or white. Terry Winograd and Jonathan Grudin have separately described the rival communities of scientists and engineers that emerged from that early work. Both men have explored the challenge of fusing the two contradictory approaches. In particular, in 2009 Winograd set out to build a Program on Liberation Technology at Stanford to find ways that computing technologies could improve governance, enfranchise the poor, support human rights, and implement economic development, along with a host of other aims.

Of course, there are limits to this technology. Winograd makes the case that whether computing technologies are deployed to extend human capabilities or to replace them is more a consequence of the particular economic system in which they are created and used than anything inherent in the technologies themselves. In a capitalist economy, if artificial intelligence technologies improve to the point that they can replace new kinds of white-collar and professional workers, they will inevitably be used in that way. That lesson carries forward in the differing approaches of the software engineers, AI researchers, roboticists, and hackers who are the designers of these future systems. It should be obvious that Bill Joy's warning that "the future doesn't need us" is just one possible outcome. It is equally apparent that the world transformed by these technologies doesn't have to play out catastrophically.

A little over a century ago, Thorstein Veblen wrote an influential critique of the turn-of-the-century industrial world, *The Engineers and the Price System*. He argued that, because of the power and influence of industrial technology, political power would flow to engineers, who could parlay their deep knowledge of technology into control of the emerging industrial economy. It certainly didn't work out that way. Veblen was speaking to the Progressive Era, looking for a middle ground between Marxism and capitalism. Perhaps his timing was off, but his basic point, as echoed a half century later at the dawn of the computer era by Norbert Wiener, may yet prove correct. Today, the engineers who are designing the artificial intelligence–based programs and robots will have tremendous influence over how we will use them. As computer systems are woven more deeply into the fabric of everyday life, the tension between augmentation and artificial intelligence has become increasingly salient.

What began as a paradox for me has a simple answer. The solution to the contradiction inherent in AI versus IA lies in the very human decisions of engineers and scientists like Bill Duvall, Tom Gruber, Adam Cheyer, Terry Winograd, and Gary Bradski, who all have intentionally chosen human-centered design.

At the dawn of the computing age, Wiener had a clear sense of the significance of the relationship between humans and their creations—smart machines. He recognized the benefits of automation in eliminating human drudgery, but he also worried that the same technology might subjugate humanity. The intervening decades have only sharpened the dichotomy he first identified.

This is about us, about humans and the kind of world we will create.

It's not about the machines.

ACKNOWLEDGMENTS

After reporting on Silicon Valley since 1976, in 2010 I left that beat at the *New York Times* and moved to the paper's science section. The events and ideas in this book have their roots in two series that I participated in at the paper while I reported on robotics and artificial intelligence. "Smarter Than You Think" appeared during 2010 and "The iEconomy" in 2012.

Glenn Kramon, who has worked with me as an editor since we were both at the *San Francisco Examiner* in the mid-eighties, coined the "Smarter Than You Think" rubric. I am a reporter who values good editors, and Glenn is one of the best.

The case I made to the paper's editors in 2010 and the one I describe here is that just as personal computing and the Internet have transformed the world during the past four decades, artificial intelligence and robotics will have an even larger impact during the next several. Despite the fact that our machines are increasingly mimicking our physical and intellectual capabilities, they are still entirely man-made. How they are made will determine the shape of our world.

Gregg Zachary and I have been both competitors and collaborators for decades, and he remains a close friend with an encyclopedic knowledge of the impact of technology on society. John Kelley, Michael Schrage, and Paul Saffo are also friends who have each had innumerable conversations with me about the shape and consequences of future computing technologies. I have for years had similar conversations with Randy Komisar, Tony Fadell, and Steve Woodward on long bike rides. Jerry Kaplan, who has returned to the world of artificial intelligence after a long hiatus, has real insight into the way it will change the modern world.

John Brockman, Max Brockman, and Katinka Matson are more than wonderful agents; they are good friends. At Harper-Collins my editor, Hilary Redmon, understood that if my last book borrowed its title from a song this one should come from a poem. Her colleague Emma Janaskie was tremendously helpful in navigating all the details that go into producing a book.

Special thanks to Iris Litt and Margaret Levi, who as directors of the Center for Advanced Study in Behavioral Studies at Stanford University, allowed me to join the community of social scientists in the hills overlooking Silicon Valley. Thanks also to Phil Taubman for introducing me to the Center.

When I was unable to obtain a visa to report in China in 2012, John Dulchinos pointed me to Drachten and my first factory of the future. In my reporting travels Frank Levy and David Mindell at the Massachusetts Institute of Technology took time to discuss the effects of robotics on the workplace and the economy. Larry Smarr, director of the California Institute for Telecommunications and Information Technology, has frequently hosted me and is always a decade or two ahead in seeing where computing is heading. Mark Stahlman was generous in offering insights on Norbert Wiener and his impact.

Mark Seiden, whose real-world computing experience stretches back to the first interactive computers, took time away from his work to help with editing, offering technical insight. Anders Fernstedt delved into the archives for gems from Norbert Wiener that had been lost for far too long. He painstakingly went through several of my drafts, offering context and grammar tips.

Finally, to Leslie Terzian Markoff for sharing it all with me.

NOTES

In cases where quotes are not attributed, they are based on the author's interviews.

PREFACE

1. This distinction was famously made by Richard Stallman, an iconoclastic software developer who pioneered the concept of freely shared software.

1 | BETWEEN HUMAN AND MACHINE

1. John Markoff, *What the Dormouse Said: How the Sixties Counterculture Shaped the Personal Computer Industry* (New York: Viking, 2005), 282.
2. Moshe Y. Vardi, "The Consequences of Machine Intelligence," *Atlantic,* October 25, 2012, http://www.theatlantic.com/technology/archive/2012/10/the-consequences-of-machine-intelligence/264066.
3. Frank Levy and Richard J. Murnane, "Dancing with Robots: Human Skills for Computerized Work," http://content.thirdway.org/publications/714/Dancing-With-Robots.pdf.
4. J. C. R. Licklider, "Man-Computer Symbiosis," *IRE Transactions on Human Factors in Electronics* HFE-1 (March 1960): 4–11, http://groups.csail.mit.edu/medg/people/psz/Licklider.html.
5. John Markoff, "Can Machines Think? Humans Match Wits," *New York Times,* November 9, 1991, http://www.nytimes.com/1991/11/09/us/can-machines-think-humans-match-wits.html.
6. Jonathan Grudin, "AI and HCI: Two Fields Divided by a Common

Focus," *AI Magazine,* Winter 2009, http://research.microsoft.com/apps/pubs/default.aspx?id=138574.

7. John McCarthy, book review of B. P. Bloomfield, *The Question of Artificial Intelligence: Philosophical and Sociological Perspectives,* in *Annals of the History of Computing* 10, no. 3 (1988): 224–229.

8. Kevin Kelly, *Out of Control: The Rise of Neo-Biological Civilization* (New York: Perseus, 1994), 33–34.

2 | A CRASH IN THE DESERT

1. Jerry Kaplan, presentation at Stanford University Probabilistic AI lunch meeting, May 6, 2013.

2. Defense Science Board, "The Role of Autonomy in DoD Systems," U.S. Department of Defense, July 2012, http://www.acq.osd.mil/dsb/reports/AutonomyReport.pdf.

3. James R. Hagerty, "A Roboticist's Trip from Mines to the Moon," *Wall Street Journal,* July 2, 2011, http://www.wsj.com/articles/SB10001424052702304569504576405671616928518.

4. John Markoff, "The Creature That Lives in Pittsburgh," *New York Times,* April 21, 1991, http://www.nytimes.com/1991/04/21/business/the-creature-that-lives-in-pittsburgh.html.

5. John Markoff, "Google Cars Drive Themselves, in Traffic," *New York Times,* October 9, 2010, http://www.nytimes.com/2010/10/10/science/10google.html?pagewanted=all.

6. "Electronic Stability Control Systems for Heavy Vehicles," National Highway Traffic Safety Administration, 2012, http://www.nhtsa.gov/Laws+&+Regulations/Electronic+Stability+Control+(ESC).

7. John Markoff, "Police, Pedestrians and the Social Ballet of Merging: The Real Challenges for Self-Driving Cars," *New York Times,* May 29, 2014, http://bits.blogs.nytimes.com/2014/05/29/police-bicyclists-and-pedestrians-the-real-challenges-for-self-driving-cars/?_php=true&_type=blogs&_r=0.

8. Lawrence D. Burns, William C. Jordan, and Bonnie A. Scarborough, "Transforming Personal Mobility," The Earth Institute, Columbia University, January 27, 2013, http://sustainablemobility.ei.columbia.edu/files/2012/12/Transforming-Personal-Mobility-Jan-27-20132.pdf.

9. William Grimes, "Philippa Foot, Renowned Philosopher, Dies at 90," *New York Times,* October 9, 2010, http://www.nytimes.com/2010/10/10/us/10foot.html.

10. "Transportation and Material Moving Occupations," Occupational Outlook Handbook, Bureau of Labor Statistics, http://www.bls.gov/ooh/transportation-and-material-moving/home.htm.
11. Vannevar Bush, "As We May Think," *Atlantic Monthly,* July 1, 1945, http://www.theatlantic.com/magazine/archive/1945/07/as-we-may-think/303881.
12. Peter Norvig, keynote address, NASA Innovative Advanced Concepts Conference, Stanford, California, February 5, 2014.

3 | A TOUGH YEAR FOR THE HUMAN RACE

1. John Markoff, "Skilled Work, without the Worker," *New York Times,* August 18, 2012, http://www.nytimes.com/2012/08/19/business/new-wave-of-adept-robots-is-changing-global-industry.html.
2. Ibid.
3. Norbert Wiener, *Collected Works with Commentaries,* ed. Pesi Masani (Cambridge, MA: MIT Press, 1985), 272.
4. "Father of Cybernetics Norbert Wiener's Letter to UAW President Walter Reuther," August 13, 1949, https://libcom.org/history/father-cybernetics-norbert-wieners-letter-uaw-president-walter-reuther.
5. Flo Conway and Jim Siegelman, *Dark Hero of the Information Age: In Search of Norbert Wiener, The Father of Cybernetics,* Kindle ed. (New York: Basic Books, 2009), Kindle location 246.
6. Anthony Carew, *Walter Reuther* (Manchester, UK: Manchester University Press, 1993).
7. Conway, *Dark Hero of the Information Age,* 246.
8. Stephen Meyer, "'An Economic "Frankenstein"': UAW Workers' Response to Automation at the Ford Brook Park Plant in the 1950s," *Michigan Historical Review* 28 (2002): 63–90.
9. "Wiener Denounces Devices 'For War': M.I.T. Mathematician Rebuffs Bid to Harvard Symposium of Calculating Machinery," *New York Times,* January 9, 1947.
10. Norbert Wiener, "A Scientist Rebels," *Atlantic Monthly,* January 1947.
11. John Markoff, "In 1949, He Imagined an Age of Robots," *New York Times,* May 20, 2013, http://www.nytimes.com/2013/05/21/science/mit-scholars-1949-essay-on-machine-age-is-found.html?pagewanted=all.
12. Ibid.
13. Ibid.

14. Carew, *Walter Reuther,* 144.
15. The Ad Hoc Committee on the Triple Revolution, "The Triple Revolution," *Liberation,* April 1964, http://www.educationand democracy.org/FSCfiles/C_CC2a_TripleRevolution.htm.
16. Mark D. Stahlman, "Wiener's Genius Project" (invited paper, IEEE 2014 Conference on Norbert Wiener in the 21st Century, 2014).
17. Steve J. Heims, *John von Neumann and Norbert Wiener: From Mathematics to the Technologies of Life and Death* (Cambridge, MA: MIT Press, 1980), 343.
18. Norbert Wiener, *God and Golem, Inc.: A Comment on Certain Points where Cybernetics Impinges on Religion* (Cambridge, MA: MIT Press, 1964), 29.
19. "Machines Smarter Than Men? Interview with Dr. Norbert Wiener, Noted Scientist," *U.S. News & World Report,* February 24, 1964, http://21stcenturywiener.org/wp-content/uploads/2013/11/Machines-Smarter-Than-Man-Interview-with-Norbert-Wiener.pdf.
20. Defense Science Board, "The Role of Autonomy in DoD Systems," U.S. Department of Defense, July 2012, http://www.acq.osd.mil/dsb/reports/AutonomyReport.pdf.
21. John Maynard Keynes, "Economic Possibilities for Our Grandchildren," in *Essays in Persuasion* (New York: W.W. Norton & Co, 1963), 358–373.
22. Jeremy Rifkin, *The End of Work: The Decline of the Global Labor Force and the Dawn of the Post-Market Era* (New York: Putnam, 1995), xvii.
23. John Markoff, "Armies of Expensive Lawyers, Replaced by Cheaper Software," *New York Times,* March 4, 2011, http://www.nytimes.com/2011/03/05/science/05legal.html?pagewanted=all.
24. Erik Brynjolfsson and Andrew McAfee, *Race Against the Machine* (Lexington, MA: Digital Frontier Press, 2011).
25. Paul Beaudry, David A. Green, and Ben Sand, "The Great Reversal in the Demand for Skill and Cognitive Tasks," NBER Working Paper No. 18901, National Bureau of Economic Research, March 2013, http://www.economics.ubc.ca/files/2013/05/pdf_paper_paul-beaudry-great-reversal.pdf.
26. Ibid.
27. James Manyika, Susan Lund, Byron Auguste, and Sreenivas Ramaswamy, "Help Wanted: The Future of Work in Advanced

Economies," McKinsey Global Institute, March 2012, http://www.mckinsey.com/insights/employment_and_growth/future_of_work_in_advanced_economies.

28. Robin Harding, "US Has Lost 2M Clerical Jobs since 2007," *Financial Times,* April 1, 2013, http://www.ft.com/intl/cm/s/0/37666e6c-9ae5-11e2-b982-00144feabdc0.html#axzz3V2czZqsP.

29. Melody Johnson, "Right-Wing Media Attack Obama for Accurate Remarks on Business' [*sic*] Investment in Automated Machines," MediaMatters for America, June 15, 2011, http://mediamatters.org/research/2011/06/15/right-wing-media-attack-obama-for-accurate-rema/180602.

30. "Are ATMs Stealing Jobs?" *Economist,* June 15, 2011, http://www.economist.com/blogs/democracyinamerica/2011/06/technology-and-unemployment.

31. Ben Sumers, "Bank Teller Case Study" (unpublished, 2012).

32. Erik Brynjolfsson and Andrew McAfee, *The Second Machine Age: Work, Progress, and Prosperity in a Time of Brilliant Technologies* (New York: W. W. Norton & Co., 2014), 127.

33. Jaron Lanier, *Who Owns the Future?* Kindle ed. (New York: Simon & Schuster, 2014), Kindle location 222–230.

34. Tim O'Reilly, Google+, January 9, 2014, https://plus.google.com/+TimOReilly/posts/F85gaWoBp3Z.

35. Matthieu Pélissié du Rausas, James Manyika, Eric Hazan, Jacques Bughin, Michael Chui, and Rémi Said, "Internet Matters: The Net's Sweeping Impact on Growth, Jobs, and Prosperity," McKinsey Global Institute, May 2011, http://www.mckinsey.com/insights/high_tech_telecoms_internet/internet_matters.

36. "The Last Kodak Moment?" *Economist,* January 12, 2012, http://www.economist.com/node/21542796.

37. Ray Kurzweil, *The Singularity Is Near: When Humans Transcend Biology* (New York: Penguin Books, 2006).

38. "The Coming Technological Singularity: How to Survive in the Post-Human Era," VISION-21 Symposium, NASA Lewis Research Center, NASA technical reports, NASA CP-10129, March 30–31, 1993, https://www-rohan.sdsu.edu/faculty/vinge/misc/singularity.html.

39. Robert Geraci, *Apocalyptic AI: Visions of Heaven in Robotics, Artificial Intelligence, and Virtual Reality,* reprint edition (New York: Oxford University Press, 2012).

40. Moshe Y. Vardi, "The Consequences of Machine Intelligence,"

Atlantic, October 25, 2012, http://www.theatlantic.com/tech nology/archive/2012/10/the-consequences-of-machine-intelli gence/264066.

41. Moshe Y. Vardi, "If Machines Are Capable of Doing Almost Any Work Humans Can Do, What Will Humans Do?" (white paper presented at Innovation 4 Jobs Conference, Menlo Park, California, 2013).

42. Author telephone interview with Moshe Vardi, May 9, 2013.

43. "IFR: Robots Improve Manufacturing Success & Create Jobs," press release, International Federation of Robotics, February 28, 2013, http://www.ifr.org/news/ifr-press-release/ifr-robots-im prove-manufacturing-success-create-jobs-469.

44. Robert J. Gordon, "Is US Economic Growth Over? Faltering Innovation Confronts the Six Headwinds," *Policy Insight* 63 (September 2012), Centre for Economic Policy Research, http://www .cepr.org/sites/default/files/policy_insights/PolicyInsight63.pdf.

45. Robert J. Gordon, "The Demise of U.S. Economic Growth: Restatement, Rebuttal, and Reflections," NBER Working Paper No. 19895, February 2014, National Bureau of Economic Research, http://www.nber.org/papers/w19895.

46. "Robert Gordon, Erik Brynjolfsson Debate the Future of Work at TED 2013," TED Blog video, April 23, 2013, https://www.you tube.com/watch?v=ofWK5WglgiI.

47. Robert J. Gordon, "Why Innovation Won't Save Us," *Wall Street Journal,* December 21, 2012, http://online.wsj.com/news/articles /SB10001424127887324461604578191781756437940.

48. Gordon, "The Demise of U.S. Economic Growth."

49. Craig Trudell, Yukiko Hagiwara, and Jie Ma, "Humans Replacing Robots Herald Toyota's Vision of Future," *BloombergBusiness,* April 7, 2014, http://www.bloomberg.com/news/2014-04-06/ humans-replacing-robots-herald-toyota-s-vision-of-future.html.

50. Stewart Brand, "We Are As Gods," *Whole Earth Catalog*, Fall 1968, http://www.wholeearth.com/issue/1010/article/195/ we.are.as.gods.

51. Amir Efrati, "Google Beat Facebook for DeepMind, Creates Ethics Board," *Information,* January 27, 2014, https://www.the information.com/google-beat-facebook-for-deepmind-creates-ethics-board.

52. "Foxconn Chairman Likens His Workforce to Animals," *WantChina Times,* January 19, 2012, http://www.wantchinatimes .com/news-subclass-cnt.aspx?id=20120119000111&cid=1102.

53. "World Population Ageing 2013," Department of Economic and Social Affairs Population Division, (New York: United Nations, 2013) http://www.un.org/en/development/desa/population/pub lications/pdf/ageing/WorldPopulationAgeing2013.pdf.

54. "Robot Caregivers Help the Elderly," Digital Agenda for Europe, A Europe 2020 Initiative, May 5, 2014, http://ec.europa.eu/ digital-agenda/en/news/robot-caregivers-help-elderly.

55. Jeffrey S. Passel and D'vera Cohn, "U.S. Population Projections: 2005–2050," Social & Demographic Trends, Pew Research Center, February 11, 2008, http://www.pewsocialtrends.org/2008/02 /11/us-population-projections-2005-2050.

56. "Baby Boomers Retire," Pew Research Center, December 29, 2010 http://www.pewresearch.org/daily-number/baby-boomers -retire.

4 | THE RISE, FALL, AND RESURRECTION OF AI

1. David C. Brock, "How William Shockley's Robot Dream Helped Launch Silicon Valley," *IEEE Spectrum,* November 29, 2013, http:// spectrum.ieee.org/at-work/innovation/how-william-shockleys-robot-dream-helped-launch-silicon-valley.

2. David C. Brock, "From Automation to Silicon Valley: The Automation Movement of the 1950s, Arnold Beckman, and William Shockley," *History and Technology* 28, no. 4 (2012), http:// www.tandfonline.com/doi/abs/10.1080/07341512.2012.756236# .VQTPKCbHi_A.

3. Ibid.

4. John Markoff, "Robotic Vehicles Race, but Innovation Wins," *New York Times,* September 14, 2005.

5. John Markoff, *What the Dormouse Said: How the Sixties Counterculture Shaped the Personal Computer Industry* (New York: Viking, 2005).

6. Pamela McCorduck, *Machines Who Think: A Personal Inquiry into the History and Prospects of Artificial Intelligence,* 2nd ed. (Natick, MA: A K Peters/CRC Press, 2004), 268.

7. Ibid., 273.

8. Brad Darrach, "Meet Shaky [*sic*], the First Electronic Person: The Fascinating and Fearsome Reality of a Machine with a Mind of Its Own," *Life,* November 1970.

9. Charles A. Rosen, "Robots, Productivity, and Quality," SRI, Artificial Intelligence Center, technical note no. 66, May 1972.

10. Nils J. Nilsson, *The Quest for Artificial Intelligence*, 2010, 77, web version, http://ai.stanford.edu/~nilsson/QAI/qai.pdf.
11. Margaret Boden, *Mind as Machine* (Oxford, UK: Oxford University Press, 2006), 381.
12. AI@50, Dartmouth College Artificial Intelligence Conference, July 13–15, 2006.
13. Boden, *Mind as Machine*, 381.
14. John McCarthy, book review of B. P. Bloomfield, *The Question of Artificial Intelligence: Philosophical and Sociological Perspectives*, in *Annals of the History of Computing* 10, no. 3 (1998).
15. Ibid.
16. Nilsson, *The Quest for Artificial Intelligence*, 77.
17. AI@50, Dartmouth College Artificial Intelligence Conference, July 13–15, 2006.
18. Interview with John McCarthy, Stanford University, July 19, 2001.
19. Steven Levy, *Hackers: Heroes of the Computer Revolution* (New York: Anchor Press/Doubleday, 1984).
20. Interview with John McCarthy, Stanford University, July 19, 2001.
21. Raj Reddy, "Celebration of John McCarthy's Accomplishments," Stanford University, March 25, 2012, YouTube, https://www.you tube.com/watch?v=J_QGryGFb2o.
22. Arthur L. Norberg, "An Interview with Bruce G. Buchanan," June 11–12, 1991, Charles Babbage Institute, University of Minnesota, http://conservancy.umn.edu/bitstream/107165/1/oh230bb.pdf.
23. Hans Moravec, *Mind Children: The Future of Robot and Human Intelligence* (Cambridge, MA: Harvard University Press, 1988), 20.
24. John McCarthy, "A Proposal for the Dartmouth Summer Research Project on Artificial Intelligence," August 31, 1955, http://www-formal.stanford.edu/jmc/history/dartmouth/dartmouth.html.
25. William J. Broad, "Computer Scientists Stymied in Their Quest to Match Human Vision," *New York Times*, September 25, 1984, http://www.nytimes.com/1984/09/25/science/computer-scientists-stymied-in-their-quest-to-match-human-vision.html.
26. John McCarthy, "Programs with Common Sense," Stanford University, 1959, http://www-formal.stanford.edu/jmc/mcc59.pdf.
27. "The Dynabook of Alan Kay," History of Computers, http://history-computer.com/ModernComputer/Personal/Dynabook.html.
28. Robert Geraci, *Apocalyptic AI: Visions of Heaven in Robotics,*

Artificial Intelligence, and Virtual Reality, reprint edition (New York: Oxford University Press, 2012), 2.

29. John Markoff, "John McCarthy, 84, Dies; Computer Design Pioneer," *New York Times,* http://www.nytimes.com/2011/10/26/science/26mccarthy.html?pagewanted=all.

30. Hans Moravec, "Today's Computers, Intelligent Machines and Our Future," Stanford University, July 21, 1976, http://www.frc.ri.cmu.edu/~hpm/project.archive/general.articles/1978/analog.1978.html.

31. Hans Moravec, "The Role of Raw Power in Intelligence," May 12, 1976, http://www.frc.ri.cmu.edu/~hpm/project.archive/general.articles/1975/Raw.Power.html.

32. Moravec, "Today's Computers, Intelligent Machines and Our Future."

33. Markoff, *What the Dormouse Said.*

34. Les Earnest, "Stanford Cart," December 2012, http://www.stanford.edu/~learnest/cart.htm.

35. Ibid.

36. Sheldon Breiner, "The Background Behind the First Airport Gun Detector," http://breiner.com/sheldon/papers/First%20Gun%20Detector%20for%20Airport--Public%20Security.pdf.

37. Robert Reinhold, "Reasoning Ability of Experts Is Codified for Computer Use," *New York Times,* March 29, 1984.

38. Jonathan Grudin, "AI and HCI: Two Fields Divided by a Common Focus," *AI Magazine,* Winter 2009, http://research.microsoft.com/apps/pubs/default.aspx?id=138574.

39. Daniel Crevier, *AI: The Tumultuous History of the Search for Artificial Intelligence* (New York: Basic Books, 1993), 203.

40. Edward Edelson, "Expert Systems—Computers That Think Like People," *Popular Science,* September 1982, 58.

41. Ibid.

42. Ibid.

43. United Press International, "New Navy Device Learns by Doing," *New York Times,* July 7, 1958.

44. John Markoff, "Researchers Announce Breakthrough in Content Recognition Software," *New York Times,* November 17, 2014, http://www.nytimes.com/2014/11/18/science/researchers-announce-breakthrough-in-content-recognition-software.html.

45. Alex Rubinsteyn and Sergey Feldman, "NIPS and the Zuckerberg Visit," *Explain My Data,* December 11, 2013 http://blog.explainmydata.com/2013/12/nips-and-zuckerberg-visit.html.

5 | WALKING AWAY

1. Manfred E. Clynes and Nathan S. Kline, "Cyborgs and Space," *Astronautics,* September, 1960, http://partners.nytimes.com/library/cyber/surf/022697surf-cyborg.html.
2. David A. Mindell, *Digital Apollo: Human and Machine in Space-flight,* Kindle ed. (Cambridge, MA: MIT Press, 2011), Kindle location 1850.
3. Ibid., Kindle location 352–364.
4. Craig Covault, "Space Leaders Work to Replace Lunar Base with Manned Asteroid Missions," *Aviation Week & Space Technology,* January 19, 2008, http://www.spaceflightnow.com/news/n0801/18avweek.
5. Slide deck courtesy of Edward Feigenbaum.
6. Morten Thanning Vendelo, "From Artificial Intelligence to Human Computer Interaction—An Interview with Terry Winograd," Association for Information Systems, *SIGSEMIS Bulletin* 2, no. 3 & 4 (2005), http://www.researchgate.net/publication/236015807_From_Artificial_Intelligence_to_Human_Computer_Interaction_-_an_interview_with_Terry_Winograd.
7. Terry Winograd lecture, "Filling in the 'H' in HCI," mediaX 2013 Conference, January 8, 2013.
8. John Markoff, "Joseph Weizenbaum, Famed Programmer, Is Dead at 85," *New York Times,* March 13, 2008, http://www.nytimes.com/2008/03/13/world/europe/13weizenbaum.html?_r=0.
9. David W. Dunlap, "Looking Back: 1978—'Farewell, Etaoin Shrdlu,'" Times Insider, *New York Times,* November 13, 2014.
10. Terry Winograd, "Procedures as a Representation for Data in a Computer Program for Understanding Natural Language," MIT AI Technical Report 235, February 1971, 38–39, http://hci.stanford.edu/winograd/shrdlu/AITR-235.pdf.
11. Hubert Dreyfus, "Alchemy and Artificial Intelligence," RAND Corporation, 1965, http://www.rand.org/pubs/papers/P3244.html.
12. Hubert Dreyfus, "Why Heideggerian AI Failed and How Fixing It Would Require Making It More Heideggerian," http://leidlmair.at/doc/WhyHeideggerianAIFailed.pdf.
13. Dreyfus, "Alchemy and Artificial Intelligence."
14. Seymour Papert, "The Artificial Intelligence of Hubert L. Dreyfus: A Budget of Fallacies," Massachusetts Institute of Technology, Project Mac, Memo. No. 154, January 1968, https://dspace.mit.edu/handle/1721.1/6084.

15. Vendelo, "From Artificial Intelligence to Human Computer Interaction."
16. Roger C. Schank and Robert P. Abelson, *Scripts, Plans, Goals, and Understanding: An Inquiry into Human Knowledge Structures,* Artificial Intelligence Series (Hillsdale, NJ: Lawrence Erlbaum Associates, 1977).
17. Searle presentation before CS22 Introduction to the Philosophy and History of Artificial Intelligence, Stanford, October 25, 2013.
18. Arthur L. Norberg, "An Interview with Terry Allen Winograd," December 11, 1991, Charles Babbage Institute, University of Minnesota, http://conservancy.umn.edu/bitstream/107717/1/oh237taw.
19. Sergey Brin, Rajeev Motwani, Lawrence Page, and Terry Winograd, "What Can You Do with a Web in Your Pocket?" *IEEE Data Engineering Bulletin* 21, no. 2 (1998): 37–47, http://citeseerx.ist.psu.edu/viewdoc/download?doi=10.1.1.36.2806&rep=rep1&type=pdf.
20. Terry Winograd, "Shifting Viewpoints: Artificial Intelligence and Human-Computer Interaction," *Artificial Intelligence* 170 (2006): 1256–1258, http://hci.stanford.edu/winograd/papers/ai-hci.pdf.
21. Jeffrey Pfeffer and Robert I. Sutton, "Knowing 'What' to Do Is Not Enough: Turning Knowledge into Action," *California Management Review* 42, no. 1 (1999).
22. Ben Shneiderman, "List of Influences: Ben Shneiderman," eagereyes, December 16, 2011, http://eagereyes.org/influences/ben-shneiderman.
23. Tandy Trower, "A Parting Salute to Cliff Nass—Social Interface Pioneer and Good Friend," Hoaloha Robotics, November 19, 2013, http://blog.hoaloharobotics.com/2013/11/19/a-parting-salute-to-cliff-nass-social-interface-pioneer-and-good-friend.
24. Tandy Trower, "Bob and Beyond: A Microsoft Insider Remembers," *Technologizer,* March 29, 2010, http://www.technologizer.com/2010/03/29/bob-and-beyond-a-microsoft-insider-remembers.
25. Nicholas Negroponte, *Being Digital* (New York: Vintage, 1996), 101.
26. Ben Shneiderman and Pattie Maes, "Direct Manipulation vs. Interface Agents: Excerpts from Debates at IUI 97 and CHI 97," *Association for Computing Machinery Interactions,* November-December 1997, http://ritter.ist.psu.edu/misc/dirk-files/Papers/HRI-papers/User%20interface%20design%20issues/Direct%20manipulation%20vs.%20interface%20agents.pdf.
27. Ibid.

6 | COLLABORATION

1. Rodney Brooks, *Flesh and Machines: How Robots Will Change Us* (New York: Pantheon, 2002), 28.
2. Ibid., 29.
3. Ibid., 31.
4. Rodney Brooks, "Elephants Don't Play Chess," *Robotics and Autonomous Systems* 6 (1990): 3–15, people.csail.mit.edu/brooks/papers/elephants.ps.Z.
5. Ibid.
6. Brooks, *Flesh and Machines,* 31.
7. Steven Levy, *Hackers: Heroes of the Computer Revolution* (New York: Anchor Press/Doubleday, 1984), 132.
8. R. H. MacMillan, *Automation: Friend or Foe,* (Cambridge, UK: Cambridge University Press, 1956), 1.
9. Levy, *Hackers,* 130.
10. Lee Felsenstein, "The Golemic Approach," LeeFelsenstein.com, http://www.leefelsenstein.com/wp-content/uploads/2013/01/Golemic_Approach_MS.pdf.
11. Ibid., 4.
12. Evgeny Morozov, "Making It," *New Yorker,* January 13, 2014, http://www.newyorker.com/arts/critics/atlarge/2014/01/13/140113crat_atlarge_morozov?printable=true&currentPage=all.
13. "Lee Felsenstein and the Convivial Computer," *Convivial Tools,* July 23, 2007, http://www.conviviality.ouvaton.org/spip.php?article39.
14. Lee Felsenstein, "The Tom Swift Terminal; or, A Convivial Cybernetic Device," LeeFelsenstein.com, http://www.leefelsenstein.com/wp-content/uploads/2013/01/TST_scan_150.pdf.
15. Dennis Liu, "Office 2010: The Movie," YouTube, July 9, 2009, http://www.youtube.com/watch?v=VUawhjxLS2I.
16. Steven Sinofsky, "PM at Microsoft," Steven Sinofsky's Microsoft TechTalk, December 16, 2005, http://blogs.msdn.com/b/techtalk/archive/2005/12/16/504872.aspx.
17. "The Lumiere project: The Origins and Science Behind Microsoft's Office Assistant," *Robotics Zeitgeist,* 2009, http://robotzeitgeist.com/2009/08/lumiere-project-origins-and-science.html.
18. Lexi Krock, "The Final Eight Minutes," Nova, WGBH, October 17, 2006, http://www.pbs.org/wgbh/nova/space/final-eight-minutes.html.
19. Jessica Marshall, "Victory for Crowdsourced Biomolecule

Design," *Nature,* January 22, 2012, http://www.nature.com/news/
victory-for-crowdsourced-biomolecule-design-1.9872.

20. Sherry Turkle, *Alone Together: Why We Expect More from Technology and Less from Each Other* (New York: Basic Books, 2011),
101.

21. Miriam Steffens, "Slaves That Are Reflections of Ourselves,"
Sydney Morning Herald, November 19, 2012, http://www.smh
.com.au/small-business/growing/slaves-that-are-reflections-of-
ourselves-20121118-29k63.html.

22. Daniel Crevier, *AI: The Tumultuous History of the Search for
Artificial Intelligence* (New York: Basic Books, 1993), 58.

23. Ken Jennings, "My Puny Human Brain," *Slate,* February 16, 2011.

7 | TO THE RESCUE

1. Stuart Nathan, "Marc Raibert of Boston Dynamics," *Engineer,*
February, 22, 2010, http://www.theengineer.co.uk/in-depth/
interviews/marc-raibert-of-boston-dynamics/1001065.article
#ixzz2pevPQoYI.

2. Public Information Office, "Fact Sheet on 'Simon,'" Columbia
University, May 18, 1950, http://www.blinkenlights.com/classic
cmp/berkeley/simonfaq.html.

3. Ivan E. Sutherland, letter to Grey Walter, November 10, 1957, http://
cyberneticzoo.com/cyberneticanimals/1956-mechanical-animal-
willian-robert-bert-sutherland-ivan-e-sutherland-american.

4. Daniel Lovering, "Radioactive Robot: The Machines That Cleaned
Up Three Mile Island," *Scientific American,* March 27, 2009, http://
www.scientificamerican.com/article/three-mile-island-robots.

5. John Markoff, "The Creature That Lives in Pittsburgh," *New York
Times,* April 21, 1991, http://www.nytimes.com/1991/04/21/bus
iness/the-creature-that-lives-in-pittsburgh.html?src=pm&page
wanted=3&pagewanted=all.

6. Marvin Minsky, "Telepresence: A Manifesto," *IEEE Spectrum,*
August 31, 2010, http://spectrum.ieee.org/robotics/artificial-
intelligence/telepresence-a-manifesto.

7. Ibid.

8. Gill A. Pratt, "Robot to the Rescue," *Bulletin of the Atomic Scientists,* December 3, 2013, http://thebulletin.org/robot-rescue.

9. "2013 DARPA Robotic Challenge Trials Overview," Defense
Advanced Research Projects Agency, http://archive.darpa.mil/
roboticschallengetrialsarchive.

10. Singularity Hub, "Industrial Perception Robot Arm Box Toss," YouTube, April 17, 2013, http://www.youtube.com/watch?v=IX xU4sL6hq0.

11. "Guidelines for Scientific Research," Graduate School of Information Science and Technology, University of Tokyo, March 2011, http://www.i.u-tokyo.ac.jp/edu/others/pdf/guideline_en.pdf.

12. Charlie Rose, "Amazon Unveils Futuristic Plan: Delivery by Drone," *60 Minutes Overtime,* December 1, 2013, http://www.cbs news.com/news/amazon-unveils-futuristic-plan-delivery-by-drone.

13. DARPAtv, "Darpa's Pet-Proto Robot Navigates Obstacles," Defense Advanced Research Projects Agency, October 24, 2012, https://www.youtube.com/watch?v=FFGfq0pRczY.

14. Hans Moravec, *Mind Children: The Future of Robot and Human Intelligence* (Cambridge, MA: Harvard University Press, 1988), 15.

15. Eric Berger and Keenan Wyrobek, "PR1 Robot Cleans a Room," Salisbury Robotics Lab, Stanford University, 2007, YouTube, https://www.youtube.com/watch?v=jJ4XtyMoxIA.

16. "The Hopkins Beast," Field Robotics Center, 1960, http://www.frc .ri.cmu.edu/~hpm/talks/revo.slides/1960.html.

8 | "ONE LAST THING"

1. John Markoff, "A Free and Simple Computer Link," *New York Times,* December 8, 1993, http://www.nytimes.com/1993/12/08/business/business-technology-a-free-and-simple-computer-link.html?pagewanted=all&src=pm.

2. Danielle Newnham, *Mad Men of Mobile: Leading Entrepreneurs and Innovators Share Their Stories, from SIRI to SHAZAM,* Kindle ed. (Seattle: CreateSpace, 2013), Kindle location 129.

3. Ibid., Kindle location 137–139.

4. John Sculley and John A. Byrne, *Odyssey: Pepsi to Apple: A Journey of Adventure, Ideas, and the Future* (New York: HarperCollins, 1987).

5. David Greelish, "An Interview with Computing Pioneer Alan Kay," *Time,* April 2, 2013, http://techland.time.com/2013/04/02/an-interview-with-computing-pioneer-alan-kay.

6. Gordon E. Moore, "Cramming More Components onto Integrated Circuits," *Electronics* 38, no. 8 (1965), 114–117.

7. Paul Pangaro, "'Getting Started' Guide to Cybernetics," Pangaro .com, http://www.pangaro.com/definition-cybernetics.html.

8. J. J. O'Connor and E. F. Robertson, "Andrew Gordon Speedie Pask," School of Mathematics and Statistics, University of St. Andrews, Scotland, http://turnbull.mcs.st-and.ac.uk/~history/Biographies/Pask.html.

9 | MASTERS, SLAVES, OR PARTNERS?

1. Claire Cain Miller, "Technology, Aided by Recession, Is Polarizing the Work World," *New York Times,* July 22, 2014, http://www.nytimes.com/2014/07/23/upshot/how-technology-aided-by-recession-is-transforming-the-work-world.html?rref=upshot.
2. Bureau of Labor Statistics, "Occupational Employment Projections to 2022," December 2013, http://www.bls.gov/opub/mlr/2013/article/occupational-employment-projections-to-2022.htm.
3. Zeynep Tufekci, "Failing the Third Machine Age: When Robots Come for Grandma," *Medium,* July 22, 2014, https://medium.com/message/failing-the-third-machine-age-1883e647ba74.
4. Louise Aronson, "The Future of Robot Caregivers," *New York Times,* July 19, 2014, http://www.nytimes.com/2014/07/20/opinion/sunday/the-future-of-robot-caregivers.html.
5. Ken Goldberg, "Robots with Their Heads in the Clouds," Aspen Ideas Festival 2014, *Medium,* August 25, 2014, https://medium.com/aspen-ideas/robots-with-their-heads-in-the-clouds-e88ac44def8a.
6. "Asimov's Three Laws of Robotics + the Zeroth Law," March 1942, Jeremy Norman's HistoryofInformation.com, http://www.historyofinformation.com/expanded.php?id=4108.
7. Defense Science Board, "The Role of Autonomy in DoD Systems," U.S. Department of Defense, July 2012, http://www.acq.osd.mil/dsb/reports/AutonomyReport.pdf.
8. John Markoff, "Already Anticipating 'Terminator' Ethics," *New York Times,* November 24, 2013, http://www.nytimes.com/2013/11/24/science/already-anticipating-terminator-ethics.html.
9. Bill Joy, "Why the Future Doesn't Need Us," *Wired,* April 2000, http://www.wired.com/wired/archive/8.04/joy.html.
10. Eric Horvitz and Bart Selman, "Interim Report from the Panel Chairs," AAAI Presidential Panel on Long-Term AI Futures, Association for the Advancement of Artificial Intelligence, August 2009, http://research.microsoft.com/en-us/um/people/horvitz/note_from_AAAI_panel_chairs.pdf.
11. Mark Prigg, "Google Sets Up Artificial Intelligence Ethics Board to

Curb the Rise of the Robots," *Daily Mail,* January 29, 2014, http://www.dailymail.co.uk/sciencetech/article-2548355/Google-sets-artificial-intelligence-ethics-board-curb-rise-robots.html.

12. *Blade Runner,* directed by Ridley Scott (1982; Burbank, CA: Warner Bros.).

13. Leon Neyfakh, "Do Our Brains Pay a Price for GPS?" *Boston Globe,* August 18, 2013, http://www.bostonglobe.com/ideas/2013/08/17/our-brains-pay-price-for-gps/d2Tnvo4hiWjuybid5UhQVO/story.html.

INDEX

NOTE: Page references in *italics* refer to photos.

JOHN MARKOFF has been a technology and science reporter at the *New York Times* since 1988. He was part of the team of *Times* reporters that won the 2013 Pulitzer Prize for Explanatory Reporting and is the author of *What the Dormouse Said: How the Sixties Counterculture Shaped the Personal Computer Industry.* He lives in San Francisco, California.